Biology Run Amok!

ALSO BY MARK C. GLASSY

Movie Monsters in Scale: A Modeler's Gallery of Science Fiction and Horror Figures and Dioramas (McFarland, 2013)

The Biology of Science Fiction Cinema (McFarland, 2001; softcover 2006)

Biology Run Amok!

The Life Science Lessons of Science Fiction Cinema

MARK C. GLASSY

Foreword by DENNIS DRUKTENIS

McFarland & Company, Inc., Publishers
Jefferson, North Carolina

LIBRARY OF CONGRESS CATALOGUING-IN-PUBLICATION DATA

Names: Glassy, Mark C., 1952– author.
Title: Biology run amok! : the life science lessons of science fiction cinema / Mark C. Glassy ; foreword by Dennis Druktenis.
Description: Jefferson, North Carolina : McFarland & Company, Inc., Publishers, 2018 | Includes bibliographical references and index.
Identifiers: LCCN 2017049324 | ISBN 9781476664729 (softcover : acid free paper) ♾
Subjects: LCSH: Science fiction films—History and criticism. | Biology in motion pictures. | Realism in motion pictures.
Classification: LCC PN1995.9.S26 G57 2018 | DDC 791.43/615—dc23
LC record available at https://lccn.loc.gov/2017049324

BRITISH LIBRARY CATALOGUING DATA ARE AVAILABLE

ISBN (print) 978-1-4766-6472-9
ISBN (ebook) 978-1-4766-2592-8

© 2018 Mark C. Glassy. All rights reserved

No part of this book may be reproduced or transmitted in any form or by any means, electronic or mechanical, including photocopying or recording, or by any information storage and retrieval system, without permission in writing from the publisher.

On the cover: The Bride of Frankenstein, 1935 (Universal Pictures/Photofest); background Laboratory © 2018 CSA-Images/iStock

Printed in the United States of America

McFarland & Company, Inc., Publishers
Box 611, Jefferson, North Carolina 28640
www.mcfarlandpub.com

For Donna
my wife, my life

Table of Contents

Foreword by Dennis Druktenis   1
Introduction   3

Laboratories in Science Fiction Films   9
   The Microscope   10
   Dr. Van Helsing's Experiment   28

Frankenstein   31
   The Notebooks of Frankenstein   32
   The Chalk Notes of Dr. Gustav Niemann   42
   The Laboratory of Dr. Septimus Pretorius   49
   The Spark of Life: The Popular Science of Mrs. Mary Shelley   66
   Boris Karloff: The Walking FrankenDead   77

Physiology   91
   The Hairy Who Are Scary   92
   Drugs in Science Fiction Cinema   111
   Hormones, the Scariest of Them All!   126
   Invasion of the Microbes   147

Surgery   167
   Brains, Craniums, and Heads, Oh My!   168

Genetics and DNA   177
   The Legacy of Doctor Moreau   178

Population Biology   185
   Foods of the Gods   186

Radiation Biology   201
   Amazing Colossal Science   202

The Nurse in Science Fiction Films   219
   History of Nursing   221

Appendix: The Films   239
Scary Monsters *Article Bibliography*   241
General Bibliography   242
Index   243

Foreword
by Dennis Druktenis

Mark C. Glassy's first article for *Scary Monsters Magazine* appeared in 2009 in our *Scary Monsters 2009 Yearbook, Monster Memories*, #17. I figured why make Mark's *Scary* writing debut start with only one article and why not feature two, so not *one* but *two* articles were featured in MM #17. The *Scary Monsters 2009 Yearbook* was quick to sell out for many scientific reasons (supply and demand?), but perhaps it was because of Mark's two debut articles! So you better quickly purchase not *one* but *two* copies of this book before it sells out. You can save both or perhaps give one to a friend.

Mark became a regular writer a few issues later with *Scary Monsters* issue #72 in 2009.

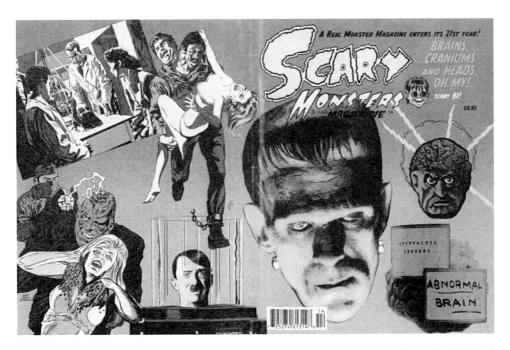

Cover to *Scary Monsters* #81 featuring the article "Brains, Craniums, and Heads, Oh My!"
©2016 Dennis Druktenis Publishing & Mail Order, Inc.

Foreword by Dennis Druktenis

A number of his pieces over the years have been Rondo Award nominees for Best Article.

It never ceases to amaze me that monster movies fans from around the country, and even the world, share many of the same Monster Memories, which turned them into the fans they are today. Perhaps Dr. Glassy can explain all this in more scientific terms. Actually, he did tackle that very topic back in *Scary Monsters* #81, 2012, with "Brains, Craniums, and Heads, Oh My!"

From 2009, *Scary Monster* readers had a nonstop fun learning experience compliments of Mark all the way through issue #99 in 2015. Now these articles are gathered together, so everyone can enjoy them all in a single tome.

Submitted for your approval is the Skinny on all things Scary. Enjoy!

Dennis J. Druktenis is the publisher, editor-in-chief and creator of *Scary Monsters Magazine*.

Introduction

Film is a powerful tool for teaching. After all, the most dominant art form is film and more people see film than all other forms of art or entertainment combined. As such, movies have had a significant influence on the general public and have sparked many popular trends. Also, films are an excellent yard stick in which to measure "the sign of the times" since each film is forever locked into its view of the world at the time of its making. With the popularity of film it is no surprise that movies can also be used as a means to teach and educate the public. For many, much of the "big picture" of science has been obtained from popular media. The first time people saw a skeleton or a human brain was either in a film or a cartoon, often years before a classroom setting was involved. These images have significant impacts on young impressionable minds.

The current generation, one I think could be called the "Jurassic Park Generation," has primarily obtained their science from popular culture and media. How many people out there really think that dinosaurs can be cloned from surviving cells in insects encased in amber? Film does influence and often times the pseudo-science presented in science fiction (SF) films affects the decision making of people, oftentimes thinking that what they see on the screen is factual. Whether consciously or subconsciously film does make an impact. One of the goals of writing the essays assembled in this book is to help the reader better understand the difference between science and pseudo-science as a presented in SF films. This is important as more and more of the audience becomes familiar with basic science principles and their applications.

After all, even the acronym, "DNA" is being used in standard television, radio, and print commercials ("it's in our DNA") so the general population does understand the concept of DNA and that these three capitalized letters somehow, magically, control our lives. DNA is the universal genetic language for all species from unicellular bacteria all the way up to humans and everything in between. (And since there is yet no definite proof of extraterrestrials—so this is therefore speculation—I am quite willing to predict that, when we find it [or it finds us], life out there is also based on DNA.)

For this particular book the films of interest are SF films. Though obvious, a key word in science fiction is "science" which is something each film's brain trust (writer,

Introduction

producer, director) can take at face value, twist in novel ways, completely ignore, or completely make out of whole cloth. This is OK and is highly dependent upon the script and, of course, the budget. The other obvious key word in SF is "fiction" meaning going anywhere and everywhere so anything can, and probably will, happen. The chapters in this book are not intended as criticism but rather an embracing of the genre, looking at both the science and the fiction in SF cinema, as a teaching tool.

Professionally, I am a cancer immunologist. As a faculty member at the University of California, San Diego (UCSD) I taught a class on antibodies, proteins of the immune system. Though I enjoyed teaching very much over the years I never had any more than 25 students in each class. In having a positive teaching experience at UCSD, I kept thinking what it would be like to have a class of hundreds, perhaps thousands, maybe even 10,000 students. A problem that could be solved through writing.

One of the reasons for writing these essays was to engage the general public in a discourse on science (the teacher in me). Most people (some estimated 95 percent) get their science outside of the classroom and, unfortunately, not all of it is accurate, and SF films do play a part in this. This is the above referenced, Jurassic Park Generation, who get most of their science from popular media. Hopefully, this book will help them understand and appreciate science more clearly. This could raise the collective "civic science literacy" of the general public and certainly enrich their lives. In essence, raise society's science IQ. By raising this collective bar the general public may be better informed in future decision making, job skills, and an overall general higher standard of intellectual living. Current issues such as global warming, stem cells, ozone hole, artificial life, biosensors, and cloning will be essential elements of future life and with an elevated sense of science literacy the general public will be better served and be able to make intelligent decisions.

To continue, all of these essays appeared in the magazine *Scary Monsters* (SM). The overall purpose of writing these articles, essays that educate and entertain, was to help *SM* magazine readers explore the personal relevance of science and integrate scientific knowledge into complex practical solutions, which should help them focus on authentic problems. Hopefully, these essays help *SM* readers develop a better understanding of the social and institutional basis of scientific credibility (science knowledge should empower those to make reasonable judgements about the trustworthiness and validation of scientific claims). Furthermore, these essays may help *SM* readers build on their own enduring, science-related interests by fostering the development of idiosyncratic interests, a habitual curiosity, and lifelong science-related hobbies. In addition, they help focus on real problems, demonstrate a multi-disciplinary approach, and help create a culture of meaningful experimentation. In the simplest of terms, as a teacher, SF film is my textbook and *SM* readers my classroom. Through the readership of *SM* magazine, I now have that dreamed-of vision of 10,000 students.

This book is primarily designed for those intelligent lay individuals, like the readership of *Scary Monsters* magazine, who are interested in cinema in general and SF

Introduction

cinema in particular. This book will shed some light on what it would really take to actually create some of the SF film monsters we all love and hate. It is certainly not as simple as SF cinema makes it out to be. In biology there are countless ways things can go wrong and that goes a long way to explain why just about every SF monster goes awry because things do indeed go wrong, terribly wrong.

Being a Baby Boomer Monster Kid (born in 1952) I grew up on SF films and read all of the monster magazines, in particular, *Famous Monsters of Filmland*, which significantly contributed to my passion. As I became a professional scientist, the Monster Kid in me thrived, and thanks to my education and training, I began to appreciate SF films on a whole new level. To enjoy movies in general, it is advised that the viewer should assume a "willing suspension of disbelief," and when applied to the watching of SF films with their varying degrees of scientific credibility, the advisement changes to "leave your brain at the door," which I do, but I can do that while maintaining the critical eye of a scientist (maybe I only leave part of my brain at the door…). It is the dual appreciation as a Monster Kid and working scientist that drives my passion for using SF films to teach biology. I am proud to have the moniker of Monster Kid Scientist.

Science in Hollywood

It is very important to realize that the art of storytelling may be at odds with scientific accuracy. In science fiction films plot often trumps science, but science can also improve the storyline. As film audiences become more sophisticated, the genre has as well, bringing more verisimilitude to SF films. Everyone wins here. It can be easy to play the role of "science accuracy police," but it's not necessary, or even beneficial. Science fiction movies aren't created as documentaries, so 100% scientific accuracy isn't a reasonable expectation. A key element in evaluating SF films is to consider the public's general understanding of science when the film was produced. In other words, in terms of SF film realism what science is known and how much of this is generally understood by the public?

Each subsequent decade's audience became more knowledgeable, sophisticated and critical, so to keep up, script writers incorporated contemporary science to drive movie plots. To paraphrase Arthur C. Clarke, the science in 21st century SF films would seem like magic to early 20th century film audiences.

Science fiction films from the first half of the 20th century focused on glands and bubbling liquids in a myriad of glass containers because those were the familiar trends and symbols of science. As the atomic era began after World War II, the SF films of the 1950s focused on radiation and kept pace with the public understanding (i.e., fears) of the "problems of radiation." As medicine advanced from the 1960s into the current DNA-age, filmmakers continued to integrate contemporary scientific concepts into the plots of their movies. As science knowledge progressed then SF film plots kept pace with the science.

Introduction

Since SF films are something many are familiar with a common bond is shared by the film going audience that can be tapped into. It is this collective bond, a universal familiarity, that can be exploited and serve as an educational textbook. It is this collective consciousness of the film going audience, irrespective of the diverse types of people involved, that can be utilized. For example, just about everyone is familiar with the general plot of the film, *Frankenstein*, so a common knowledge is collectively shared uncoupled from who the person is. With such a common shared interest this can then be used as a teaching tool for a wide and diverse audience (i.e., classroom). With this in mind, in the right context, entertainment can be educational as long as it is not overly obvious. And in turn, educational material can be very entertaining if done in the right way. Science education should be focused on helping people use science in daily life instead of emphasizing knowledge and skills.

An important question to ask is does accurate science make a film better? This is certainly debatable and depends upon the story and budget. As Robert A. Heinlein said in regard to the film version of his book, *Destination Moon*, "realism is expensive."

Furthermore, is science accuracy worth the expense? This too is debatable. Film is a form of entertainment and most movie science is just that, entertainment but it is not without merit. It is clear that story drives SF film plots and not necessarily science. Film is a visual medium and the language, look, and symbols of science don't always translate well on screen. This naturally requires the tweaking of details, resulting in the "fudging" of scientific fact, making "folk science," or pseudo-science. Since SF films are made to entertain, and most deliver in a big way, if it requires sacrificing a degree of scientific accuracy, then so be it.

And for me personally, speaking as a Monster Kid Scientist, I am OK with some bending of the rules of science because I see film medium as a form of entertainment. A SF film does not have to be 100 percent accurate to be entertaining. When I do the proverbial "check your brain at the door" when watching SF films then was I entertained by the film irrespective of its science accuracy? More often than not the answer is yes, though levels of entertainment vary considerably.

Another quality of SF film that warrants evaluation is the level of scientific sincerity. If the science is flawed but the film feels sincere then it is better appreciated by the audience. Any variation of scientific accuracy and scientific sincerity in SF cinema helps make the impossible seem plausible. On top of all this are purposeful mistakes and/or accidental mistakes by Hollywood science. Some of this can be attributed to budgets and the skill level of those who made the film.

If you love SF films as much as I do then the good, the bad, and the downright ugly are all of interest and all have some sort of merit. Overall, the films collectively discussed in all these essays do indeed cover the good, the bad, and the ugly which is something of interest because instead of looking at each film on its own merit (or lack thereof) the oeuvre of SF films was looked at with a different assessment, namely the science involved. This provides a completely different perspective to SF films and helps to segment these films into different categories that deserve a closer look.

Introduction

All of these chapters first appeared in the magazine *Scary Monsters* magazine. The readership of *SM* is quite intelligent and it is this intelligent audience (read: classroom) that caught my attention.

With the "educate and entertain" mantra in mind, the original articles in *Scary Monsters* were written in a light and familiar tone so as not to off-put interested readers with pedantic descriptions of seemingly complex biology. A goal here was to make biology fun and entertaining and using SF films as a textbook helped to meet that goal.

One particular theme that has been revisited several times in these articles is the laboratory sets in SF films. These sets, where the science action takes place, are interesting windows within the film production to analyze and understand a significant amount of biology. One important question is whether the various lab sets were themselves pertinent to the work at hand as presented in the films. Were the lab sets adequate for the work or was this bench bling just for show without any real purpose other than it looks cool on a lab bench in a film? What is on these lab benches does provide much insight about the type and nature of the supposedly offered science in these films.

Another major theme is body physiology, both *in*side and *on* the body. Our bodies are the vesicles we use to carry out our lives. Though much of our body physiology is determined by genetics, a good portion of it comes from lifestyle and diet. Through proper lifestyle and diet then many genetic deficiencies can by overcome or at least mitigated. And after all, many of our favorite screen monsters are derived from humans (and human parts) so human physiology is pertinent.

Due to the diverse nature of the themes of the articles plus the fact that they were written years apart meant that there is some overlap in ideas and quotes to discuss certain points. Since these articles are presented as a single book all at once then some overlap is inevitable. Some of the film quote examples can be used in many ways, each indicative of the context in which they were used. For example, films like *The Amazing Colossal Man* and *House of Dracula* are referenced several times. It is important to retain the integrity of each of the original articles without too much mixing and blending together of the essays so the articles are presented as stand alones. If the articles were edited and all mixed together then the intent of the original articles would be lost. So please forgive this author if some material in each of the articles contains a wee bit of overlap. Which brings me up to my final comment. Each of the original articles ended with the same appellation and the same applies here too so I will close this introduction: "Thank you for reading. It's back to the lab for me. Stay healthy and eat right."

Laboratories in Science Fiction Films

It would be difficult to come up with an accurate number of hours I have spent in a laboratory in my life. Over 40 plus years as a professional scientist I easily have spent over 100,000 total hours in a lab, averaging 10 hours a day (scientists by and large happily work many hours). And that's not counting evenings and weekends of which there were plenty. And during that 40 years, the type, nature, and style of lab equipment, what I call "bench bling," has significantly changed. In the 21st century just about all equipment is run by computers, but back in the 1970s when I began my formal training, computers were not yet fully integrated into to our work environments. Much of the work back then was done by hand and now many procedures are in kit form or automated. Since so much has changed in the bench bling over the years, we can refer to the laboratory set in a SF film as a historical marker to identify when the movie was made. In movies, just like cars, clothing, and technology reveal their era, the bench bling present in a SF movie serves to do the same. While watching these films as a Monster Kid Scientist, the lab sets stood out much more to me than what may have been intended. While some SF laboratories are offbeat, some are quite serious and elaborate, and others are just down right laughable, they are all entertaining.

1

The Microscope

I was raised in a medical family. My father was a pathologist and I spent a lot of time as a youth gazing down a microscope in his office. During 6th grade science I was allowed to bring in one of my dad's old microscopes to class and we all looked at the amazing microbes in pond scum. The microscope is a tool of the trade for a biomedical scientist and I have used them all in one form or another, from the smallest pocket microscope to the high-tech transmission electron microscope.

A cowboy has his horse, the cop has his gun, and our inveterate SF scientist has his trusty microscope. They are one of the most dramatic set pieces and when we, the audience, see a cinemascientist gazing into a microscope, we get an immediate sense that something interesting is being observed, something that will prove to be pivotal to the plot. In most cases we just see the scientist looking in a microscope and his reaction right afterwards. On a few occasions we actually get a glimpse of what was seen by the scientist. What is shown in these scenes has varied from drawings, to real images of cells and tissues, often bearing no relation to the life form supposedly being viewed.

Many of us were first exposed to both telescopes and microscopes while watching movies, probably way before such instruments were first seen in a formal setting, such as a school classroom. The function of these instruments are easy to grasp. The telescope is used for seeing distant objects, invisible to the naked eye, and the microscope is used for seeing tiny objects, also invisible to the naked eye.

The sophistication of the microscope in SF films varies dramatically from embarrassingly simple devices (a kiddie scope) to some astonishingly expensive equipment used only in the most sophisticated research labs. In some instances, the value of these high-end microscopes is more than the entire budget of the film. For example, a large speciality microscope called a fluorescence microscope appeared in the film *Frozen Alive* (1964), and the price of that single piece of equipment surpassed many film budgets. And in the film, *War of the Gargantuas* (1966), we see scientists working with an even more expensive electron microscope. Such high dollar microscopes would not be brought to a film set. The crew would film on location at the research lab housing such an instrument, and the film would gain credibility by showcasing sophisticated equipment in its authentic environment.

1. The Microscope

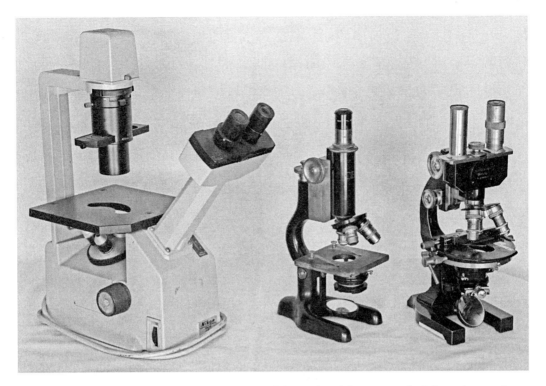

Three light microscopes from the collection of the author. The one on the left, an inverted version with the objective lenses underneath the stage, was marketed during the 1970s. Its permanent electrical light source, located on top, shines directly down and through the optics (note the electrical cord wrapped around the base). The microscope in the middle, a monocular version with three objective lenses, is from the 1920s or 1930s and quite a popular style in the Golden Age of Cinema. The mirror underneath its stage is purposely pointed almost straight up as often seen in our favorite SF films and as such there is no way enough incident light can get to the optics for proper viewing. The microscope on the right, a binocular version, also with three objective lenses, is from the 1940s and is seen in many films of the time. Its mirror is pointed at the right angle to get incident light.

A Light History of Microscopes

Long ago in ancient times someone picked up a piece of glass (molten sand) that was thicker in the middle than the edges and when looking through it noticed that objects appear larger. The lens was invented. They were named lenses because they were shaped like the seeds of a lentil. In Roman times such lenses were used to focus the rays of the sun causing fabrics and other materials to burst into flames.

A microscope (from the Greek: μικρός, *mikrós*, "small" and σκοπεῖν, *skopeîn*, "to look" or "see") is an instrument used to see objects that are too small for the naked eye. Microscopes provide a window into the cellular and molecular world through the use of a lens or combination of lenses. They provide access to the fascinating worlds within worlds and invite humans to contemplate the wonders of life beyond

Cover to *Scary Monsters* #90 featuring the article "The Microscope in Science Fiction Films." ©2016 Dennis Druktenis Publishing & Mail Order, Inc.

what is visible. There are many types of microscopes, the first invented, was the optical microscope which uses light to see the sample. Other types include the electron microscope (two versions, transmission electron microscope and scanning electron microscope) and various types of scanning probe microscopes. Confocal microscopes are a type of fluorescence microscope and are related to optical microscopes.

The most common type of microscope is the optical or light microscope. This is an instrument that contains one or more lenses that create an enlarged image of a sample. These optical microscopes use refractive glass to focus the incoming illuminating light into the eye. There are two major types of light microscopes and they are distinguished by the eye pieces. A monocular microscope has a single eyepiece to look through and a binocular microscope has two, one for each eye. Binocular microscopes became more prominent during the early 20th century. Typical magnifications of light microscopes are up to 1500x with a theoretical limit of 200 nanometers due to the limited resolution of diffracted light. These light microscopes, even one with perfect lenses and illumination, can not distinguish objects that are smaller than half the wavelength of light. White light has an average wavelength of 0.55 micrometers so half of that is 0.275 micrometers. Any two objects that are closer together than 0.275 micrometers will not be distinguishable and blur. To see objects smaller than 0.275 micrometers a different source of "illumination," one with a shorter

1. The Microscope

wavelength than light, is necessary. For cellular imaging the maximal resolution for light microscopes is about 10 nanometers. Shorter wavelengths of light, such as ultraviolet, are one way to improve resolution. Current instruments allow the resolution of tens of nanometers. As we move into the 21st century there are continuing improvements in light sources, cameras, detectors, labeling technology, computers, and image analysis software. Signal-to-noise ratios have been improved and now 3-D imaging of intact cells is possible. Needless to say microscopy has come a long way.

Microscopic Development

Though an earlier version was allegedly made in 1590 in the Netherlands, two eyeglass makers, Hans Lippershey and Zacharias Janssen are often credited as being the first inventors of the optical light microscope. They experimented with several lenses in a tube and discovered that objects were greatly enlarged when two convex lenses were combined. Since then, microscopy has enabled highly efficient and accurate molecular, genetic, and cellular imaging for countless research and clinical applications.

Giovanni Faber coined the name "microscope" for Galileo Galilei's compound microscope in 1625 (Galileo called it the "occhiolino" or "little eye"). The earliest tube microscope was merely a tube with a plate for the object at one end and at the other a lens which magnified objects about 10 times their actual size. Galileo worked out the principles of lenses and made a significant improvement with the ability to focus the lenses.

The father of microbiology, Anton van Leeuwenhoek (1632–1723), began as an apprentice in a dry goods store and used magnifying glasses to count the threads in cloth. He taught himself how to grind and polish new lenses that resulted in magnifications of up to 270x. With such lenses Leeuwenhoek was able to build microscopes that lead to the discoveries he is known for. He was the first to describe single celled organisms such as bacteria or yeast, to glimpse the amazing amount of tiny life teeming in a single drop of water, or blood moving through capillaries. He called the small microorganism life forms he first observed under a microscope, "animalcules." For the record, on October 9, 1676, Leeuwenhoek reported the discovery of his animalcules, the first window into the much larger microbial world, to the Royal Society of London.

The English father of microscopy is Robert Hooke, who not only confirmed Leeuwenhoek's discoveries but also significantly improved on the design of the light microscope by describing how to make single-lens versions. After Hooke few improvements were made in microscopes until the middle of the 19th Century when several companies began to manufacture fine optical instruments with magnifications up to 1250x. The level of magnification depended on how precisely the lenses were ground during production.

In 1644 the first detailed drawing of living tissue, a fly's eye, was rendered based on observations made with the use of a microscope. During the 1660s and 1670s the microscope was extensively used in research and intimate drawings of miniscule biological structures became popular. Scientific illustrators had a huge impact on influencing public interest in biology. Their work inspired subsequent generations of scientists to gaze into a microscope to further explore Nature's invisible wonders.

Since 1647 when Leeuwenhoek first observed cells in a microscope he built, imaging has been central to studies of the molecules and organisms that make up the microscopic world. During the past 366 years, we have definitely come a long way from those first observations with the introduction of new technologies including the electron microscope and super resolution microscopy in the early 1980s, both of which have vastly increased magnification and resolution and enabled imaging at single nanometer resolutions of the same image.

Types of Microscopes

For every job there is the right tool. Not all microscopes are created equal and the main difference is in the optics. And the technology of microscopes is evolving at a pace similar to computers, where it seems every six months or so new technology is introduced making the previous version obsolete. Light sources are no longer light bulbs with limited hours but consist of LEDs and lasers that can last significantly longer, with higher intensities, different wavelengths, and a wider range of uses.

Not only are there many types of microscopes there are also many types of microscopy, ranging from the simple observations of cells and tissues, to observing the movements of specific molecule or protein complexes in real time, to examining the details of a cell's surface or cytoarchitecture. Each of these approaches provide different information, so the combination of two or more microscopy systems provides more refined data. This is referred to as multimodal microscopy where different systems are coordinated and correlated to provide a superior resolution of the sample. Such multimodal combinations enable scientists to observe all three-dimensions of cells and their shapes. Some current microscopy systems offer fast 3-D structured illumination microscopy, wide field microscopy, and localization microscopy techniques, all within the same system.

Microscopes can also be separated into different classes. One major class is based on what interacts with the sample to create an image, such as light (optical microscope), electrons (electron microscope) or a probe (scanning probe microscope). Another major class depends on whether the microscope analyzes the sample via a scanning point (scanning electron microscope and confocal optical microscope) or all at once (transmission electron microscope). Each class of microscope can give dramatically different versions of the same image.

A distinguishing feature of light microscopes is that they use lenses, both optical and electromagnetic, to magnify the image created when a wave of light passes through the sample or reflected by the sample. The resolution is limited by the wave-

1. *The Microscope*

length used to image the sample, the shorter the wavelength the higher the resolution. For the scanning and electron microscopes the lenses focus a spot of light or electrons onto the sample and the reflected or transmitted waves are then analyzed at a much higher resolution than that of light microscopes.

For standard light microscopes to work properly an even light source must be shined through the sample, through the optics, and into the eye for observation. Thicker samples will block more light passing through to the eye thereby preventing any meaningful observation. Typically, samples are no more than 10 microns thick (one micron is a millionth of a meter) so enough light can effectively pass through to see. It wasn't until the late 19th Century that effective illumination sources were developed that have given rise to the modern era of microscopy. This extreme even lighting overcame many of the limitations of older techniques.

Potential light sources in addition to natural light are ultraviolet, near infrared, and fluorescence. Ultraviolet light is useful to image samples transparent to the eye, near infrared light can be used to see circuitry embedded in silicon boards (silicon is transparent in near infrared light), and fluorescent light can specifically illuminate samples to allow special viewing. For phase contrast microscopy there are small phase shifts in the light passing through the sample specimen that are converted into amplitude and contrast shifts to better see the samples. Now, in the early part of the 21st century the traditional optical microscope has evolved into a digital microscope where the sample is no longer directly viewed through an eyepiece but through the sensors of a digital camera and displayed on a computer monitor.

The most recent developments in light microscopy involve not the microscope itself but rather in fluorescence microscopy, a technique where samples are labeled with fluorescent molecules, called fluorophores, so individual cellular structures can readily be visualized. For example, there are specific fluorescent labels for DNA, cellular proteins, and organelles such as the mitochondria that allow precise analysis of all cellular components in real time. These techniques allow for the analysis of cell structures both at the molecular level and whole cell level. The rise of fluorescence microscopy also drove the development of modern microscope design, such as with confocal laser scanning microscopes starting in the 1980s. Many fluorescent features are now incorporated into current microscopes to broaden their function. In the 21st century significant research is focused on developments of super resolution of fluorescently labeled samples, and early results suggest that such structured illumination can improve resolution by two to four fold.

In the early 20th century a significant alternative to traditional light microscopes was developed using electrons rather than light to generate an image. These electron microscopes work on the same principle as optical light microscopes but use electrons instead of light and electromagnetics in place of glass lenses. Since the wavelengths of electrons are much smaller than that of light the resolution of electron microscopes is much higher than traditional light microscopes and can easily reach magnifications of several hundred thousand fold. There are three main types of electron microscopes.

Laboratories in Science Fiction Films

For transmission electron microscopy electrons pass through the sample, analogous to basic optical microscopy, which are then detected whereas for scanning or scanning probe electron microscopy electrons are scattered over the surface of objects with a fine electron beam. Since electrons are strongly scattered by passing through samples careful preparation of these samples is necessary. The first transmission electron microscopes were introduced in 1931 and the first scanning electron microscopes were introduced in 1935. The first commercial transmission electron microscopes were marketed during the 1950s and the first commercial scanning electron microscope was available in 1965. In the 1980s the first scanning probe microscopes were developed and were closely followed in 1986 by the invention of the atomic force microscope.

The evolution of microscopes has co-evolved with advances in optics, light sources, and within the last generation, computers. However, in a basic analysis the glass lens of a microscope has not changed much in the last 100 years. What has dramatically changed has been major improvements in computers and sensor technology, to enhance what can be seen.

These technological developments coincide with the evolution of methods to embed and stain samples, the discovery of fluorescent proteins for intracellular labeling, and new techniques for monitoring molecular interactions within living cells.

Now digital cameras can be readily mounted on microscopes for enhanced imaging capabilities. As our understanding of biology becomes more sophisticated, microscopes keep pace with the development of more sophisticated technology. The most important properties of any microscope will depend upon the intended application, so features such as lens objectives, filters, imaging detectors, and illumination sources are important. Many modern microscopes are modular and can readily be upgraded, depending upon the application (for example, fluorescence requires special filters), to maintain top performance.

If Leeuwenhoek or Hooke were alive today they would be able to watch physiological processes in real time, observe a virus infecting a lymphocyte, a bacterium replicating in a host organism, or a bacteriophage injecting its DNA into a host cell. They would also be able to detect and precisely locate single molecules and monitor their movement over hours or days, or study the embryological development of small animals. Microscopy has become poetical as we can now see a tapestry of cells and molecules artfully woven together. Charles Darwin commented that the world seen through a microscope provides "endless forms most beautiful." Nearly 400 years after its creation, the lens of the microscope still remains the most accessible window into the cellular and molecular world.

The Films

The following films all provide a point of view perspective of what the cinema-scientist sees as he peers through his trusty microscope. The age and style of the

1. The Microscope

microscope reveals either when the film was made or when the story took place. Microscopes made during the 1800s are quite different in design from those made during the early 1900s, which are drastically different from those made in the latter half of the century. The 21st century microscope is digital and the images are viewed on a flat screen monitor and not through an eyepiece. Images can easily be scanned and available software is able to detect even the most minute details. The traditional old-school microscope will soon become a relic from the past: a museum piece or something found in only the classic science fiction films, with the current digital microscopes infiltrating global biomedical laboratories and clinics and movie screens.

Son of Frankenstein (1939)

In a key scene in the refurbished laboratory of Baron Wolf von Frankenstein (Basil Rathbone), the good doctor and his assistant, Thomas Benson (Edgar Norton), along with Ygor (Bela Lagosi) are examining the monster and take a sample of the his blood to evaluate through a microscope. The microscope is a top quality binocular instrument, very popular at the time, with three primary objective lenses. A special light source is positioned at the correct angle towards the microscope mirror adding a nice touch of realism.

The microscope POV in this scene indeed shows a blood smear, though it is impossible to tell which species of blood. Dr. von Frankenstein first warms up the microscope slide over a Bunsen burner, likely to burn off residual moisture from the bottom of the slide. The first view shows a relatively low power magnification, probably 40x, and later we see a higher power magnification, probably 100x. Though the large round cells seen floating around are red blood cells (RBCs), with an occasional

Lon Chaney with his trusty microscope from the 1922 film *A Blind Bargain*. Note the mirror underneath the microscope stage that can be angled to maximize the light entering the optics and the simple (low wattage) omni-directional light bulb as the light source. Such a weak, scattered light source would not provide enough illumination for proper viewing through the lens of this microscope.

white cell floating by too, that is not the most interesting detail. The fact that the RBCs appear to be floating suggests there is way too much fluid on the glass slide, allowing fluid flows to occur. In order to properly observe cells under a microscope the cells should be stationary and not moving.

Also noteworthy are the small "specs" that appear to be rapidly and randomly moving about, some up and down and others sideways. These little specs are several species of bacteria. What this means is the blood sample is heavily contaminated, since normal blood does not have *any* bacteria. It probably took a while for the cinematographer to set up the shot and during that time the sample became contaminated just by being exposed to the air. The high power POV shot is even more heavily contaminated than the low power POV shot, suggesting the higher magnification filming was done later, allowing more bacteria to grow, divide, and overgrow the sample.

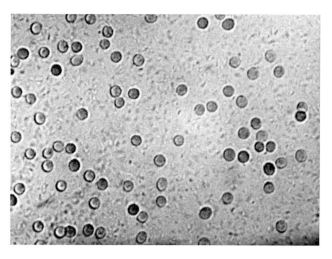

From the film *Son of Frankenstein*. Low power point-of-view image of the Frankenstein Monster's blood sample obtained by Wolf von Frankenstein. This is an actual blood sample, highly contaminated. The round cells are red blood cells whereas all those small blurry specs are contaminating bacteria.

If such a sample was actually taken from a patient, then that patient would most likely be dead from septic shock due to all the bacteria present. If we apply the willing suspension of disbelief there is room for interpretation. Maybe the monster's physiology and immune system was such that it could tolerate, and may even require, such bacterial invaders to supplement his metabolism. Or, if the monster's body was constantly in a state of bacterial shock this would naturally affect his behavior, resulting in outbursts of unrestrained mayhem.

Dr. von Frankenstein says, "I've never seen blood like that before ... polymorphocellular ... extreme hemocrosis ... the alpha leukocytes apparently do not dissolve. The entire structure of the blood is quite different from that of a normal human being." Then while looking at the higher power magnification he says, "Cells seem to be battling one another as if they had a conscious life of their own." Polymorphocellular, though an interesting sounding fifty-cent word, is a made up one. A close word, polymorphonuclear refers to white cells with a multi-lobed nucleus. Hemocrosis is a term referring to an iron storage disorder characterized by excessive intestinal adsorption of dietary iron. How the good doctor came to this conclusion by looking at a sample of the monster's blood is a mystery. There is no such thing as "alpha

1. The Microscope

leukocytes" so it is unclear what does not dissolve. (Leukocyte is another word for white blood cell.) Lastly the comment describing, "cells battling one another" would suggest the monster has a healthy immune system where cells indeed battle invading germs and other microorganisms. Or perhaps he has some sort of autoimmune disorder where his own immune system battles his own cells and tissue.

House of Dracula (1945)

We get a good idea of what Count Dracula's blood might look like when the good Dr. Franz Edlemann (Onslow Stevens) examines a sample of his blood. Through the microscopic POV we see long snake-like black "cells" that have three or four "fingered" projections on whip-like ends that seem to encircle individual cells. Also, these unique cells in Dracula's blood are antigenic since an "antiserum" was developed by Dr. Edelman, to combat the vampire's blood disorder.

The microscope used is a small version of a v-shaped instrument. Since the objective lenses of this microscope are quite tiny the magnification power is limited with a 40x lens probably being the highest (for most light microscopes a 40x objective lens would be the medium range power and not the highest). A lamp pointing directly onto the microscope mirror hints at an attempt at accuracy, which is undermined by the fact that the lamp light is not on. As seen in *Son of Frankenstein*, the microscope slide was briefly waved over a Bunsen burner to help evaporate any residual moisture on the glass slide. The POV microscope image is an actual blood smear, though a photograph of one and not a "live" shot. The RBCs are plentiful with one white cell visible, just to the right of center, the one with the multi-lobed nucleus; these cells are called polymorphonuclear leukocytes. The RBCs with the long snake-like forms that end in finger-like projections are drawn over the blood smear image.

Point-of-view of Dracula's blood as seen through a microscope in the film *House of Dracula*. The smaller round cells are red blood cells and the larger, multi-lobed cell on the right side is a white cell, called a polymorphonuclear leukocyte. The long worm-like structures with the finger-like ends are unique (and fictitious) Dracula cells.

Laboratories in Science Fiction Films

The War of the Worlds (1953)

In *The War of the Worlds* a sample of Martian blood is obtained and examined under a microscope. The nice binocular microscope was contemporary for the early 1950s. The light source pointed directly at the mirror adds authenticity; the mirror appears to be aimed at the correct angle to maximize the incident light through the optics. Of the three microscope objective lenses, the lowest power is used in this scene which would be correct for examining a large view field of a blood sample. The view of the nucleated RBCs is interesting; it is a well-done drawing and each cell has a nucleus. The nucleus areas of each cell are simply cross-hatch marks with no such examples in real life (or at least life on Earth). While examining the Martian blood sample the doctor says, "I don't remember ever seeing blood traces as anaemic as these. They may be mental giants but by our standards (i.e., the human immune system), physically, they must be very primitive."

In healthy adult humans RBCs are the only cells that do not have a nucleus and, therefore, are unable to reproduce themselves like all the other body cells. Nucleated RBCs are those that have not yet matured and are usually seen in newborn infants and some patients with anemia. Nucleated RBCs are rare in normal, healthy animals. When RBCs are first formed in our bodies they do contain a nucleus. However, soon after a stem cell changes into a RBC the nucleus is extruded resulting in a typical nucleus-free red blood cell. Without a nucleus RBCs are unable to divide and have a half-life of around 120 days. When RBCs are destroyed then new stem cells divide to create new RBCs. This normal process continues throughout the lives of mammals.

In the film, the large number of nucleated RBCs in the sample are indicative of a primitive blood (and therefore, immune) system. This strongly suggests that there are very few germs on Mars, so the native Martians have no need to develop an extensive immune system. Also, since there is very little oxygen on Mars, the natives would need to have long lasting RBCs for optimal oxygen absorption. These cells would naturally need a nucleus to control all the metabolic needs of long-lived cells. All this Martian biology ascertained from a simple microscope POV shot.

She-Devil (1957)

In the first scene we are treated to a microscopic POV drawing of a fruit fly (*Drosophila melanogaster*, to be specific), apparently through a conventional light microscope. Though these fruit flies are small, they are still too large for their whole body to be viewed with such a microscope. In reality such low-power, whole body images would require a "stereo microscope" or compound microscope, typically used to view objects such as insects, rock structures, or leaves at low magnification, around 10x–20x (one is used in the film *Mutiny in Outer Space*). With the conventional light microscope shown in this scene the magnification of the fruit fly would be quite high,

1. The Microscope

and one could readily see individual cells with such a high power magnification. In simple terms, the scale is wrong.

The microscope used in this film is contemporary with late 1950s instruments and is probably a teaching level microscope rather than a research microscope. On the plus side is a nice light source in a contemporary housing unit specifically designed to illuminate microscope mirrors. Though the POV shot we see is of a low power magnification, the microscope objective lens used in this scene is the medium power objective, again much too powerful for the image shown. A medium power objective on this microscope, probably 40x, would be strong enough to see individual cells in one of the insect's antennae.

Night of the Blood Beast (1958)

This film grants two POV glimpses through a microscope. Unfortunately, the microscope is a somewhat below average monocular version with no visible light source for the microscope mirror. The middle objective lens is used and based on the type of microscope probably offers a 40x magnification.

A spaceship crash lands back on earth and a sample of blood is taken from the seemingly dead pilot. As one of the scientists says, "Dead seven hours and his blood is still alive." After looking at the pilot's blood sample under the microscope another scientist says, "I've never seen anything quite like it. Notice the way it's fighting the others.... I've seen amoebic dominance of a cell structure before but this is completely out of proportion. Normal blood has two basic cells, the red carry oxygen and the white fight infection. But this blood has three. That third cell, the big one, that's completely foreign to any blood structure. If that bacteroid is contagious then we've all been exposed." Through this POV microscope shot we see cartoon drawings of (alien) cells moving about, including some sort of fierce looking cells with tentacles (called pseudopods) that grab (and devour?) normal, "human" cells.

Later, after the pilot has been revived and appears normal, a second blood sample is obtained and again examined under the same microscope. As the scientist then explains to the pilot as he looks down the microscope, "That's what your blood looks like now. Normal in every respect [normal for a drawing!]. Four hours ago it was populated with alien amorphic cell structures." In this second POV shot we see similarly drawn cells as in the first, but instead, in this view, that third cell, "the big one," is missing. The cells in this shot also move about (via the same crude animation) and both red and white cells are visible. The red cells are the round cells and the white cells are oblong with a cell nucleus visible. Typically, in normal blood stains, red cells far out number white cells so the relatively excessive number of white cells seen here would suggest the pilot may have a blood cancer, like leukemia, that has abnormally large numbers of white cells.

Or are those "amorphic cell structures" alien stem cells with pluripotent capability? Perhaps fertilized alien egg blastocyst cells that are nourishing themselves from the pilot's body as a result of his "pregnant condition."

Monster on the Campus (1958)

We are fortunate in this film in that we get to see two microscope images. The first one is of frozen bacteria and the second is of regular, room temperature bacteria moving about. The tiny little specs seen rapidly moving about (similar to that seen in *SoF*) are individual bacteria cells, some rod shaped and some ball shaped. Such a magnification would be several 100x magnification (400x?).

The microscope in this film is an older version with the two eye piece tubes forming a V-shape. For the first view of the microscope it is actually pointed backwards. The microscope mirror is typically opposite the person viewing allowing a light source to shine directly into the optics and in this first view the scientist is directly in front of the mirror effectively blocking any light from getting through. Also, the sides of the specimen stage of the microscope have odd flaps, the purpose of which is not entirely clear (block incident light?). A later viewing of the microscope has it pointed in the right direction with the mirror away from the person. At least with this orientation maybe some of the incident room light could have been focused on the mirror. Even so, there was no obvious light source for the microscope so the images seen would be dark.

It is unclear what the actual first POV shot is, though it is stated that the sample is "frozen bacteria." This first view down a microscope looks like simple dirt or debris often seen with dusty and unclean microscope slides. The second POV shot is much clearer in that it indeed is of normal bacteria all swimming about. (As mentioned, this shot is similar to that shown in the microscope scene from *SoF*, but without the red blood cells floating about.)

Blood of the Vampire (1958)

The Victorian-era microscope seen in this film has both optic tubes merge, V-like, just above the observing stage of the microscope. The optic tubes converge into a single objective lens, which limits the instrument because only one magnification (probably 40x) is available. Though v-shaped microscopes are quite useful, the view through such an instrument is actually a single circle, and not the two somewhat-overlapping circles, "binoculars-like," view shown in the film.

The reason for showing two independent fields in the same POV shot, as explained by the scientist/vampire, Dr. Callistratus (Donald Wolfit), is to show two different blood groupings, A and B, on two different glass slides. (Note: Callistratus correctly places a glass cover slip over the blood sample on the glass slide though almost forgets this step for the second slide.) Callistratus mentions there are differences between blood group A and blood group B and has his assistant look at this with the microscope (the left image in the POV shot is A and the right image in the

1. The Microscope

POV shot is B). The assistant confirms, "I see the difference." Quite frankly, both images seem identical to me. Lastly, such blood cells should not move around on a glass slide indicating there is too much fluid in the slide preparation.

Caltiki, the Immortal Monster (1959)

A single-celled organism (Caltiki), originating from the Mayan civilization, grows when exposed to radiation. A scientist places a small amount of Caltiki tissue onto a glass slide (without a coverslip over the sample) for examination. The microscope shown is a monocular version, with only two small objective lenses (most standard microscopes have three objective lenses: low, middle, and high power magnification) of which it appears the higher power is being used (in this case, based on the size of the POV image, is probably somewhere in the 20x–40x range). While the microscope appears to be a suboptimal choice for the work at hand, a dedicated small light source is correctly aimed at the microscope mirror and into the optics.

Professor John Fielding (John Merivale) comments, "It appears to be some sort of unknown organism, some fragment of an animal. A creature made up of one complete cell." If Caltiki were indeed a single-celled organism then looking at a piece of tissue under a microscope would be like looking at a bit of jello, an amorphous mass with no visible structures. The POV microscope shot reveals fibroblasts, a type of connective tissue cell. [Note: fibroblasts is plural, meaning there are *several* of these cell types, quickly debunking the concept that Caltiki was a single-celled organism.] At first the POV shot is out of focus, but near the end of the brief scene the focus tightens and the fibroblast cells become incontrovertibly visible.

Reptilicus (1961)

A lot of microscope action occurs in this film. The first microscope shot is a brief glimpse of one on a lab bench. This small monocular microscope has no visible light source for the mirror. The second view of a microscope is of one placed on a scientist's desk. This also is a monocular microscope with the standard three objective lenses and a dedicated light source directly attached to the mirror; the light cord is visible. For the third view of a microscope it is back at the lab bench only this time it is on the opposite side from when first seen on the bench. The first POV shot seen is a high magnification of a tissue sample from Reptilicus and from the shown field there are 11 white cells (the large ones) and at least 30 RBCs (the smaller cells). If such a field were from typical animal blood there would only be 2–4 white cells visible, not 11. That many white cells visible suggests some sort of hyper metabolic and/or immune state. If such a blood sample was taken from a human then there is the distinct possibility the person either has leukemia or lymphoma, both cancers of the white blood cells, and not a good sign. Since this sample came from Reptilicus then the beast has some interesting blood indeed.

Laboratories in Science Fiction Films

The second microscope scene takes place on a small desk of a recently hired helper. While eating his sandwich the helper is cleaning the microscope, a small monocular version with only a single objective lens, probably 10x or 20x magnification. The lab helper takes a bit of his sandwich (could be fish or chicken, not sure) and looks at it under the microscope. What are seen are small parasites that live in aqueous environments (water bound). If that is indeed what is crawling on his sandwich then he does have much to be concerned about. The real problem with this scene is the lack of an obvious light source to reflect off of the mirror below the viewing stage to send light up through the underside of the slide, through the optics, and into the eye of the viewer. With no visible light source it is difficult to understand how the images of the parasites were so easy to see.

Mutiny in Outer Space (1965)

An alien fungus from "ice caves" on the moon contaminates a space station. This fungus grows when exposed to heat (from a warm body) and is inhibited from growing by ultra cold temperatures. A small black light microscope is seen on a back bench top, a light gray compound microscope is used to view the fungus, and a third piece of equipment appears to be an electron microscope [overkill on a space station].

Compound microscopes are low power microscopes that are useful for observing large objects. Some illumination is necessary though not as much as a light microscope requires. In the film the light gray microscope is a standard compound version with V-shaped eye tubes and a single objective lens, and is used several times. Each time it is used the microscope is pointed backwards, similar to that seen in "Monster on the Campus" in that the mirror is aimed at the scientist, instead of opposite to shine on a light source. Though there is a lamp near the mirror, it is a general overhead light, and not designed specifically for a microscope. The last time this microscope is used also pointed backwards, no light source at all is visible.

The images shown of the "alien fungal strands" appear to be synthetic fibers stretched out, as there is reflective light coming from the strands. Such reflected light would not come off the surface of actual fungus. However, it is an alien fungus, so maybe alien fungi do indeed consist of reflective elements, perhaps metal particles.

Regardless, an electron microscope has no business being on a space station, since all the necessary infrastructure to make such an instrument functional would be highly impractical. For example, they require liquid nitrogen to cool down the electron beams, a tricky item to store on a space station.

War of the Gargantuas (1968)

This film features a total of five microscopes: three standard light microscopes, an inverted light microscope, and an electron microscope. All housed at the fictitious, "Toto University Laboratory: Biological Atomic Chemistry," but given the quality of

1. The Microscope

the equipment, it must have been filmed on location at an actual biomedical lab. The first light microscope is actually a low-end version that is facing the wrong direction. The microscope mirror is inexplicably facing away from the light source, a common error in such movies. The second light microscope is shown when Dr. Stewart (Russ Tamblyn) is on the phone. This one is facing the correct direction with the mirror facing the window. Though the incident light from the window would be substantial it still would not be as strong as a dedicated light source for the microscope.

A scientist is seen walking out of the room with the electron micrograph photos taken of Gargantuas' tissue sample. Though it is stated that both photos are of "hair cells" only the first photo is even close showing a cross section of cilia, considered micro hairs, usually too small to be visible to the naked eye. Such high resolution as seen from these electron micrographs was too high to have any real meaning or interpretation. The second photo is of mitochondria, small organelles within each cell that serve as a metabolic powerhouse, and definitely not a hair sample.

Later, Dr. Stewart is examining a fresh tissue sample from the green Gargantua with a contemporary top of the line research microscope, an expensive instrument, indeed. The light source, with a dimmer knob, is integrated into the housing of the microscope. The third tube coming out the top of the microscope is there for optional camera attachment. With all the expensive equipment, one would think that the movie would get at least some of the science right, but the POV shot we see through the light microscope is inaccurate. The image is a cross-section of cilia micro hairs, similar to the first electron micrograph shown earlier in the film. But we see the cilia as though through binoculars, two separate fields of view, no overlap at all, with the left view in focus and the right view starting out of focus and then becoming in focus. Another problem with the image is that it exhibits the ultra-magnification of an electron microscope and not the light microscope that Dr. Stewart is looking through, high quality though it may be. The scale is way off. The magnification is probably around 40,000x, much stronger than a standard light microscopes that may have a maximum magnification of around 1,500x.

The final microscope we see is an expensive and impressive inverted microscope and another first rate research instrument. The inverted microscope, with the lens objectives *underneath* the viewing stage, has the light source permanently fixed on top of the instrument, so it can shine directly down the optics through the observed specimen and into the objective lens. In spite of the microscopes being misused and misrepresented, they are nicely showcased in this film. Interesting to note their total value would likely surpass the budget of most commercial films of the day.

The Green Slime (1968)

A scientist places some slime blood (a very tiny sample) on a too large glass plate for viewing under an "electron microscope," which is actually just a set of binoculars paired with a simple telescope tube. For those unfamiliar with such equipment, this very cheap prop is fairly convincing in its role as an electron microscope. The small sample on the glass plate is then placed directly between the two binocular lenses, which are about 5" apart, thereby making it physically impossible to view the sample through the eyepieces. Another discrepancy is that glass slides aren't used with electron microscopes. Those samples are prepared in an entirely different manner, either embedded in resin or placed on a metal plug. Furthermore, the cells appear to be living and cells observed under an electron microscope are in a vacuum and therefore not alive.

The sample image is shown on a flat screen and is of a liver cell undergoing cellular division, meaning one cell is splitting into two daughter cells. In the film the entire process takes up just a few seconds of screen time, whereas in reality such a cell division may take up to an hour.

Finally, to maintain an electron microscope on this space station would present the same insurmountable obstacles as in the film *Mutiny in Outer Space* where an electron microscope, also on a space station, is apparently operational. In spite of these entertaining flaws, *The Green Slime* gets points for scientific sincerity if not accuracy. The sample images being shown on a large flat screen monitor are a prescient nod to current microscopy. Microscopes around the world now use such flat screens to view the images, all digital, seen through the lens of a microscope.

The Creeping Flesh (1973)

This film is overly ambitious with seven different slide preparations for microscopic viewing. For the first few slides at least the initial preparatory steps were done correctly, with a drop of the blood specimen placed on a glass slide and a second slide used to smear the drop, essentially creating a gradient from thick to thin. However, no coverslips were used in any of the preparations. The oversights intensify and the last few slide preparations were increasingly sloppy with the blood specimen just dropped onto a glass slide with minimal smearing. The end result would be a very thick, and therefore difficult to examine, blood preparation. Also worth noting, in the first slide preparation made by Prof. Hildern (Peter Cushing), the glass slide used has a white frosted end, for information, such as an identification number, to be written on. Unfortunately, such frosted slides did not come into common use until well into the 20th century and this film is supposed to take place in Victorian 19th Century!

The microscope POV shots are also less than convincing. Two images are shown of individual red blood cells (with way too much red background color) followed by circular cells with long tentacle-like hairs, almost like a sea urchin, that help propel it and capture other cells. The third image is of a mixture of the red blood cells and the tentacle-like cells. All of them are clearly fake.

The visual star is a Victorian brass microscope where both optic tubes merge, V-shaped, into a single objective lens on top of the central optic stage. This is the

1. The Microscope

same microscope seen in *Blood of the Vampire*. Each optic tube has its own focus knob. Since there is only a single objective lens then only a single magnification is available. A second microscope is seen briefly late in the film quietly resting on a shelf. This is actually a better instrument, though admittedly not as charming as the antique used throughout the film, primarily because of the multiple objective lenses available.

Through the lens of a microscope the human eye is empowered to see a variety of images from the invisible world. In SF cinema the scope of available images expands from the invisible to the imaginary world. Cells internal, external, fictional, factual, earthbound and extra-terrestrial are on full-frontal display. We get an opportunity to see what monster blood might look like directly from the veins of Frankenstein and Dracula. We get a glimpse of alien blood from Martians and beings from other, more foreign planets. And through peering at the life thrumming in these cells, unimaginable before the advent of the microscope, we become aware that even when we are isolated from other humans, or visible life forms, we are never truly alone.

2

Dr. Van Helsing's Experiment

If you look hard enough you can dependably find a bit of science lurking in just about any film. Case in point is the 1931 film, *Dracula*. In a film focusing on the supernatural (or "anti-science," if you will) there is one key scene in which science plays a significant role in plot development. A bit of "kitchen chemistry" that Van Helsing performs onscreen is routinely done in grade school science classes to demonstrate how changes in pH (acidity vs. alkalinity) can affect color. Typically, a solution of a dye, usually phenothalein, which is red in color, changes to clear when a weak acid is added, soliciting "ohs!" and "ahs!" from students and though it is a simple trick, the effect is impressive on the big screen as well.

Just about all biological tests are either quantitative or qualitative. In short, these tests either gauge an amount of something (quantitative) or a quality of something (qualitative). Both tests are useful and entire industries exist to determine and exercise the practical applications of the results. When you donate blood both quantitative and qualitative tests are done on the sample to insure that both activity (quality) and amount (quantity) of cells present meets certain criteria. The results of blood sample testing varies. Sometimes the quality is high but the quantity is low and conversely sometimes the quality is low while the quantity is high. These two extremes provide a wide range of interpretation for these biological tests and the results are easily obtained.

By looking at the sample of donated blood under a microscope one can determine if the patient is anemic or over-producing certain types of blood cells. The microscope has the obvious advantage of allowing you to see the blood sample on a cellular level. A test tube assay on the donated blood wouldn't enable cellular evaluation, but the advantage of tube assays is that they allow measurement of the sample. Such a test can measure how much iron is present in the blood, indicative of the quality of the cells and how well they can move oxygen throughout the body. The best blood tests are those that include both a qualitative and a quantitative test so both can be compared. One result should support the other.

This brings us back to Universal Studio's *Dracula*. Not only the original 1931 Bela Lugosi version, directed by Tod Browning but also the 1931 Spanish production directed by George Melford with Carlos Villar as the Count.

At the time, to maximize production costs, Universal Studios often filmed two

2. Dr. Van Helsing's Experiment

versions of a movie, one in English and a second in Spanish to capitalize on that market. While the 1931 Universal Studios' production of *Dracula*, starring Bela Lugosi, was being filmed during the day, a separate production designed for the Spanish speaking market, was concurrently being filmed, but at night, when the day crew was sleeping. The Spanish feature more or less mimicked, scene for scene, the English production. Since the script and sets were essentially the same, it is the actors and the direction which distinguishes the two films.

The unique circumstances behind the production provides an interesting opportunity to compare the scientific accuracy of the two films. In both versions, during an autopsy on Mina, one of Dracula's early victims, a physician states, "On the throat of each victim, the same two marks." These marks were noted by Dr. Van Helsing who is searching for scientific proof for the existence of vampires and he thinks he now has a valuable clue.

The scene in Dr. Van Helsing's office when he tests Renfield's blood to determine if he is a vampire is especially interesting since the final diagnosis, "Nosferatu" in the Browning version and "vampiro" in the Melford version, was arrived at by two entirely different means.

Browning's Dracula

In a scene a little over thirty minutes into the Browning version we see the laboratory office of Dr. Van Helsing (Edward Van Sloan) at the Seward Sanitarium where he examines a sample of Renfield's blood in a test tube (a qualitative test). Most likely, he is performing a gravimetric test to determine how much iron is in the blood. The primary component of our blood is the protein, hemoglobin, carried around by red blood cells. In each hemoglobin protein molecule are embedded four iron atoms and these metal atoms are heavy, making their detection simple. The iron molecules also give red blood cells their red color. Anemic people, those with low blood counts (such as vampires), would have a low gravimetric test since they have few red blood cells, the primary carrier of hemoglobin, and therefore, few iron atoms. If Renfield was a vampire, the iron in his red blood cell count would be diminished, so that is what Van Helsing's test will evaluate.

In the Browning film the test tube demonstration involves Van Helsing using a simple dye color "assay" in which a color dye is changed to clear upon the addition of a drop of a reagent (a weak acid). This is a simple colorimetric assay in which the color (probably a deep red) is changed to clear by the addition of a small amount of acid. Van Helsing is shown adding a drop of reagent using a glass volumetric cylinder and his hands are unsteady, suggesting a lack of confidence or experience. It would have been much more practical for him to use a pipette (like an eye dropper) or another test tube to add the reagent. From the perspective of a scientist, this scene seems overly staged and not very convincing.

Laboratories in Science Fiction Films

After completing the test tube assay, changing the dark color to clear, Van Helsing dramatically states, "We are dealing with the undead. Nosferatu." Quite a bold statement to make from such a simple qualitative assay, but when taken in context with Renfield's unusual behavior, especially his appetite for living creatures, in addition to the low gravimetric blood test, the result would support Van Helsing's grim diagnosis.

Interestingly the overall scene is dark and poorly lit showing very little of the good doctor's desk and office where he performs the test. On Van Helsing's desk are a microscope, a retort, a test tube rack, a glass mortar and pestle, and other glassware. From this shot we also see other well-stocked bookshelves behind the actors. No equivalent shot exists in the Spanish version. Also on the desk is a distillation tube held up by a ring stand; its purpose is not obvious based on the work at hand.

Medford's Dracula

In Melford's film our first view of Van Helsing's desk is at the 33 minute mark. His desk is mostly the same in this version, with the addition of a lit Bunsen burner and more glassware. At first sight Van Helsing (Eduardo Arozamena) appears to be shaking a test tube. He is then shown peering into a microscope. Here Van Helsing makes his vampire diagnosis with the help of a microscope (quantitative test) whereas as in the Browning version Van Helsing makes his diagnosis from a test tube assay (qualitative).

With Renfield's blood sample ("sangre") on the slide under the microscope Van Helsing shares his diagnosis: "Vampiro." His colleague counters that "vampires are not real but superstition." And Van Helsing responds, "I can prove through testing this is not a superstition but real." Nothing like a good quantitative assay to prove your point and this version is more scientifically authentic. The microscope is mightier than the test tube.

Summary

Scientifically, direct assays are always better and can be more definitive. So, the "scientific test" to determine if a vampire exists was done (indirectly) in a test tube in Browning's film and (directly) via a microscope in the Melford version. An interesting comparison and it all comes down to a simple qualitative versus a quantitative test. Even so, both Van Helsings make the same "definitive" diagnosis: "Nosferatu" or "Vampiro."

The presence of Bela Lugosi as Count Dracula went a long way to make Browning's film a part of cinema history. The Spanish version may be superior in many cinematic ways, including the more accurate portrayal of science, but the Count, played by Carlos Villar, is somewhat comical in comparison to Lugosi and as a result the film suffers, leaving the viewer to decide what counts more in their personal enjoyment of the film, a more compelling Dracula, or more convincing science.

FRANKENSTEIN

Ever since its first publication in 1818 the novel *Frankenstein*, has not been out of circulation, attesting to its historical importance in literature and some consider it the first science fiction story. As mentioned before, my favorite film monster is Frankenstein. His undeniable charm won many hearts when the 1931 Universal Studios version starring Boris Karloff was released. This then begs the question, what makes Karloff's monster so memorable? The genius of Boris Karloff was that he made the monster real by making him relatable. Someone we can sympathize with and care for, much like the blind hermit did in *Bride of Frankenstein*, who saw the monster as a "friend." Beyond the psychology of empathy, the science in the Frankenstein franchise is rich and layered and provides unique insight into some fascinating biology.

3

The Notebooks of Frankenstein

Laboratory notebooks are used all over the world and have been kept since man first learned to write and experiment (man first learned to experiment long before man began to write). The ancient Egyptians had perhaps the first notebooks, in the form of papyrus, in keeping records of what they did and what they observed. In principle notebooks are any solid surface, though typically paper, in which writings, notes, and other relevant items can be kept for future reference. And that is the key purpose of a lab notebook, to serve as a record of work that can be referred to again if necessary.

During the course of his work of creating the monster he apparently kept copious notes and this is based on the various papers and bound volumes scattered on desks and on shelves as seen in the 1931 film. We even see Dr. Henry make several notes on his work at hand. In creating the monster Dr. Henry had to make certain reagents or solutions and follow certain steps and procedures or make a special observation. No doubt all of this was documented in his notes.

Later, either Dr. Henry, a colleague such as Dr. Waldman, or an associate (Fritz?) needed to refer to the notes because they needed to repeat a step or refer to an annotation so they needed to go back to that area of the notebook and follow what was originally described. This is essentially the same way cookbooks function. You want to look up a certain recipe so you go to that area of the cookbook and follow it. Same with many experiments in that you look up your earlier procedure, or recipe, and duplicate it. And all made possible by keeping a notebook. And with family cookbooks the "secrets" are typically passed on from mother to daughter. In the case of the Frankenstein family their notebooks containing all their family secrets are passed from father to son and so on.

These notebooks of his serve several important purposes. First is a record of what was done and what resulted from what was done (this is important if a subsequent annoyed scientist [in later films of the series] wants to refer to the work and perhaps later reproduce a particular step or procedure; this could help him avoid having to re-do certain "mistakes"). Stating the obvious, all steps and procedures should be recorded in the notebook, the more extensive the better. The second purpose is using a notebook record as a source for future reports and reference. They are referred to as the record of discovery and observation and can be quite valuable as a resource.

3. *The Notebooks of Frankenstein*

Physically lab notebooks can be any type of hard surface to record the experiments to more modern electronic sources for storing information. The most common is a bound laboratory notebook in which the pages are used in consecutive order. These notebooks have either lined pages or pages marked with grids for ease of incorporating data. Please note that our inveterate annoyed scientist, including Dr. Henry, is not bound by such requirements and to him these details are seemingly trivial matters. These scientists can (and often do) use all sorts of scrap paper to record their notes that are often kept loosely or in a file, as amply demonstrated by the amount of notes scattered about Dr. Henry's lab. In the course of my own research career I have used any number of solid surfaces, including paper, to make records of the work. (Perhaps the most unusual surface I have used in my career was in writing down notes and observations on toilet paper ... but I digress.) Data obtained from various experiments can be stored in any manner of ways, either as a hard copy embedded into notebooks or kept electronically in a computer. For that matter, depending upon the mood of the annoyed scientist, he can even use crayons to record his data.

In the pharmaceutical world notebooks also serve a third purpose, namely, as a legal document to establish priority especially in the filing of patents. In the pharmaceutical industry these notebooks are not only scrutinized daily but signed off on daily by both the working scientist and his supervisor/boss. The demands are very strict with these notebooks because billions of dollars can be at stake so utmost care is taken in their accuracy. None of this, of course, is of any concern to our annoyed scientist. All he cares about is the creation of his monster and anything not related to that is superfluous and essentially a waste of his time. And, to be sure, having a notebook notarized would be of no concern to the annoyed scientist in his creation of an SF monster. And no doubt, of no concern to Dr. Henry either.

Another use of a notebook is perhaps in a legal defense should the annoyed scientist be unfortunate enough to have to deal with this. He could use his notes to either support his defense or be used by the prosecution as evidence of guilt. (Assuming, of course, that the nabbed scientist even admitted to having such notebooks that could then be used against him.)

Finally, notebooks can also be something left to offspring as either a direct inheritance (like those found in the case presented to Dr. Wolf Frankenstein in *Son of Frankenstein*) or something found (like the notebook searched for in *Frankenstein Meets the Wolfman*).

Each page of a notebook or each separate page of whatever notes are being kept should be at least dated so you know when you did the particular experiment or made the relevant observation. Each key step and each key ingredient should also be noted as well as any specific bits of experimental data, both numerical (readouts from a machine for example) and observational (the monster looked like this, acted like that, his blood pressure, etc.). When there are common steps or procedures that have been used routinely before then it is OK to reference the prior method (for example, "see page 43 of notebook 2a for cranial nerve attachment"). Other important items to put in a notebook are time (morning, afternoon, evening, and night do influence biological rhythms), temperature, and relative humidity (if applicable).

When mistakes are made it is routine to just cross one line through the error and place the correction within the immediate vicinity on the notebook page. Even though most notes are made with an ink pen, pencils are still used quite frequently. In the pharmaceutical industry only ink pens are allowed since permanent records must be kept. The annoyed scientist will use whatever is within reach, crayons if necessary (sometimes even blood). Real notebooks are neither strewn about nor individual pages crumpled and thrown into the trash as is often seen in our favorite films, mostly out of frustration by our annoyed scientist.

The integration of data into a notebook can take many shapes and forms. Various printouts can be taped, stabled, paperclipped, glued, or just loosely laid into the notebook. This also applies to photographs, drawings, or other printed items necessary to the particular experiment. Other generated data such as brain and heart rhythm printouts, DNA sequences, or various hard forms of data can just as easily be incorporated into a notebook.

Throughout the world, lab notebooks of the 21st century have become electronic and all relevant information pertaining to a particular experiment are kept in a computer. Most of the currently obtained data are in digital format that can easily be incorporated into a computer typically in a PowerPoint format and easily sent over the Internet. Also, other hard data can be readily scanned and digitized for computers. A major advantage here is the ability to access a large amount of date quickly. The annoyed scientist just wants easy access to his notes and whatever format is easy for him is what he chooses whether it be paper, computer, memory, or a combination of all the above. His goals are self-serving so this could change the traditional concept of a science notebook. His notes are for himself after all and not necessarily meant to survive the ravages of time (nor the scrutiny of patent examiners).

For our purposes, in the original film, *Frankenstein*, and the sequel, *Bride of Frankenstein*, the official "records" of the Frankenstein canon were created by the man himself, Dr. Henry Frankenstein (and later added to by Drs. Waldman and Pretorius), and it is these records that were much sought after during the subsequent film sequels. In particular, in *Frankenstein Meets the Wolf Man*, Larry Talbot goes to extreme lengths including subterfuge and lying in order to get the Frankenstein notebook "The Secret of Life and Death."

The Films

Frankenstein (1931)

Dr. Waldman's (Edward Van Sloan) office has well-stocked bookcases on every wall. On his desk we see the notes that he has been studying and an array of notebooks scattered about on the desk and on chairs and tables. Dr. Waldman was Henry

3. The Notebooks of Frankenstein

In *Frankenstein Meets the Wolf Man*, Dr. Mannering (Patric Knowles), Baroness Elsa Frankenstein (Ilona Massey), and Larry Talbot (Lon Chaney) look at the notebook *The Secret of Life and Death* by the baroness's father.

Frankenstein's mentor at medical school, so we can safely conclude that Waldman trained him in scientific note-taking.

In Dr. Frankenstein's lab on top of a wood table abutted to a wood pillar is an open bound notebook. Two similar, but closed notebooks are to the right of this and above those are three large books. Also on the table are a mortar and pestle as well as a coffee pot and a few other items, such as a reading lamp, suggesting an active lab life. This certainly reflects an actual lab where coffee pots and other amenities are kept within easy reach.

In one particular exchange, Dr. Frankenstein says to Fritz, his assistant, "Let's have one final test" showing the true spirit of a dedicated scientist. Always tests and controls and controls and tests to make doubly sure that everything is ready and in working condition. And all of the results of the various tests are documented in a notebook.

After bringing the monster to life both Frankenstein and Waldman are seen relaxing at a table with many notes and books. Frankenstein is writing notes on separate pieces of paper and not in a bound notebook, so his note taking is quite fluid. This shows that he is more concerned about the results than he is about bound notebooks of results. After the monster dramatically enters the lab he ambles by a pillar papered with various notes demonstrating a put-it-anywhere approach, which emphasizes that Dr. Frankenstein is a working scientist and not an interior decorator, all in keeping with his profession.

When Waldman is ready to dissect the monster he is seen writing in a notebook, demonstrating that a visiting scientist also keeps notes of his work. We get a brief glimpse of the notebook page where Waldman has written, "...as per injections of 5:00–900 ∂ 12³⁰. Tuesday, 730 PM note increased resistance. Necessitating stronger and more frequent injections. However, will perform dissection at once." This bound notebook is around 10"x14" and appears to be around 125 pages long. On the table are many other loose pages of notes, anatomy drawings, and open books, and various scraps of paper with scientific scribbling.

In the many sequels to the original film there is the ongoing search for Dr. Frankenstein's "records," his "diary" and "The Secret of Life and Death." Of all these notebooks and notes which one was the "holy grail" that all else sought? Which was the singular notebook that did indeed have all those Frankenstein secrets?

Bride of Frankenstein (1935)

When the assistant, Ludwig (Ted Billings), helps Dr. Frankenstein (Colin Clive) move the table holding the heart support apparatus closer to the bride for surgery, we get a clear view of a shelf near a back wall that is loaded with large notebooks and thick collections of loose papers. So, we get a glimpse of notes or "records" that prove to be important in *Frankenstein* sequels. The large notebooks dwarf those that are prized in later sequels. One possibility is that Dr. Frankenstein created an annotated notebook that summarized and highlighted all the essential data and it is this elusive single volume "record" that everyone wants. However, it should be noted that at the end of this film the lab was "blown to atoms" so the question remains, what happened to those notebooks?

Son of Frankenstein (1939)

Baron Wolf von Frankenstein (Basil Rathbone), the son of Henry Frankenstein, and also a doctor, was given a box by the town council upon his return arrival. In his home library he opens the box and discovers it contains important papers of his father's estate. From a letter Dr. von Frankenstein solemnly reads aloud, "My son. Herein you will find my faiths, my beliefs, and my unfoldments. A complete diary of my experiments, charts, and secret formulas. In short, the sum total of my knowledge. Such as it is. Perhaps you will regard my work with ridicule or even with distaste. If so, destroy these records. But if you, like me, burn with the irresistible desire to pen-

3. The Notebooks of Frankenstein

etrate the unknown then carry on. Even though the path is cruel and tortuous, carry on. Like every seeker after truth, you will be hated, blasphemed, and condemned. But, mayhap, where I have failed, you succeed. You have inherited the fortune of the Frankensteins. I trust you will not inherit their fate."

Dr. von Frankenstein brings the box of records to the lab and we get another glimpse of its contents. Inside are single leafs of paper, rolled papers, and bound notebook. In particular are individual sheets of paper that Dr. Wolf is holding that have perforated holes for placing in 3-ring binders. During his medical examination of the monster (the last appearance of Boris Karloff in this role) the doctor dictates while his assistant Thomas Benson (Edgar Norton) takes notes on a tablet of lined paper. Perhaps these notes would then later be transferred to a bound notebook. This is actually a common practice in the science lab where oftentimes brief notes are kept on hand and then later transferred with more detailed descriptions to bound notebooks.

Ultimately, all Dr. von Frankenstein contributed to the official Frankenstein records were his "Notes and Memoranda" based on the examination of the monster. Though these notes were significant, it is his father's records that were so much sought after in the subsequent sequels.

Ghost of Frankenstein (1942)

Dr. Ludwig Frankenstein (Cedric Hardwicke), a specialist in "diseases of the mind," is the son of Dr. Frankenstein and brother of Wolf and inherited the famed notebooks, which he keeps locked in a secret panel in his library. The materials are stored in a case that resembles the one his brother received at the beginning of *Son of Frankenstein*.

Ludwig opens the secret panel and blows a layer of dust from the case from which he removes two notebooks, one of which is labeled, "Diary, Baron Heinrich Frankenstein" and the other "Notes and Memoranda, Dr. Wolf Frankenstein." Both of these notebooks are large bound volumes, approximately 10" × 14" in size, and appear to be around 150 pages long, though Wolf's appears to be slightly thicker. Wolf's notebook has the word "Memoranda" in the title which suggests this is a summary volume and not the original entry notebooks he kept.

Ludwig, a prominent brain surgeon, is seen intently studying the notebooks and uses them as a guide to perform the delicate brain surgery on the monster. Later, Ludwig's daughter, Elsa (Evelyn Ankers), looks at the diary and reads from one of the first pages, "I am sure I can harness the lightning and extract its life giving power. I need but one more part to prepare the monster for my final experiment. Tonight we shall steal another body!" At this point she stops reading realizing an unpleasant truth about her grandfather. This excerpt shows that the doctor could be cavalier in his note keeping. Here he casually refers to a repeated crime and apparently does not care nor concern himself with the thought that the authorities could use such a document against him. Not to mention the insensitivity of calling his scientific triumph a "monster."

After some flashbacks we see Elsa continuing to read from another page from the same notebook, this time about half way through, and we get a glimpse of the entry, May 20th. "The body of my creature is complete. Every physical part carefully chosen and the whole assembled. Now to give him a brain." Note that here he refers to a creature instead of a monster. (Maybe, without a brain it is a creature and with a brain it is a monster.)

These comments in the notebook are really observations and certainly appropriate to put in a notebook and appear to be summary statements and not original notes. I say this because the phrase, "every physical part carefully chosen" is chock full of detail and not something easily summarized. Dr. Henry is far too meticulous a scientist to glaze over such a phrase. The obvious question is where did all the "carefully chosen" physical parts come from? How many different cadavers contributed to the creature? No information about this is in that phrase which is something that should have been entered into a notebook. (My guess is a minimum of four cadavers were used. The body proper including the head [from a gallows], two separate hands [based on the suture scars on each wrist], and finally the brain. There may have been a fifth with a contribution of a heart but there is no real suggestion of this in the film. Note that the two bullets seen in *SoF* should still be in that beating heart!)

The two pages we see Elsa reading are of note primarily because they are unlined pages whereas typical notebook pages are lined or have grids on each page. Perhaps, these notebooks are composed of loose leaf pages that were later bound into these volumes. If indeed these notebooks are summaries and not original notes then it may make sense to have these individual summary pages bound up into volumes as apparently both Dr. Henry and Dr. Wolf did (as noted above, these bound volumes could have been prepared after the fact). Somewhat related to this is an early scene in *GoF* with Dr. Ludwig sitting at his desk in his library and amongst the notes he has in front of him is a 3-ring binder holding hole-punched pages. Loose leaf pages may be the norm for the Frankensteins.

Frankenstein Meets the Wolf Man (1943)

Larry Talbot/The Wolf Man (Lon Chaney, Jr.) is desperate to rid himself of his curse and believes that the records of Dr. Henry Frankenstein will reveal the method for doing so. He tricks Baroness Elsa Frankenstein (Ilona Massey) into helping him find the diary in the castle ruins. Talbot grabs the diary out of her hands and announces excitedly, "The Secret of Life and Death!" The bound notebook is approximately 8" × 12" in size, has a ribbed spine, and appears to be about 150 pages long.

With the notebook open about half way Dr. Mannering reads, "Matter ages because it loses energy. This artificial body I have created has been charged with superhuman power so that its span of life will be extended. Its lifetime will equal the

3. The Notebooks of Frankenstein

lives of more than 100 human beings. This my creation can never perish unless [turns page of notebook and continues reading] its energies are drained off artificially by changing the poles from plus to minus." Later, he continues reading with, "energy which can not be destroyed can be transmitted." The mental wheels of an intrigued scientist are turning. The power and draw of a scientific notebook is amazing in the right hands.

After rebuilding the lab and preparing for his key experiment Dr. Mannering reads more from the notebook: "Connecting the plus poles to the minus will charge the energy output of the nervous system as by connecting the minus to the minus...." As stated above, this is the purpose of notebooks, to be able to look up previous methods and observations to repeat them. Talbot desperately wanted to get the notebook so he could learn a way to die effectively demonstrating the strong draw power notebooks have. In this case, the notebook, to Talbot, is worth more than anything else and in his eyes is his means to his end.

During the climactic smackdown between the Wolf Man and Frankenstein (now Bela Lugosi) we see the notebook on top of a piece of machinery. In the heat of battle the Wolf Man kicks the machine and the notebook falls off. Later the notebook is miraculously replaced, as if the monsters took a break from their fighting to protect the revered notebook by placing it out of harm's way.

House of Frankenstein (1944)

Dr. Gustav Niemann (Boris Karloff), wants to perfect his own experiments in brain transplantation and needs Frankenstein's notes. He bribes his hunchbacked assistant, Daniel (J. Carrol Naish) to help find them, promising him, "If I had Frankenstein's records to guide me I could give you a perfect body."

While at Neusadt Prison, Dr. Niemann uses the walls of his cell as his "notebook" and uses chalk to write his notes on the walls. At the start of the film Niemann temporarily strangles a guard until he gets his chalk not only demonstrating the resolve Niemann has to do his work but also the extremes he goes to, regardless of who gets in the way. This effectively shows that he prefers chalk over food. And by using his prison walls Niemann amply demonstrates that any surface can be used as a notebook! Unfortunately, all that work was lost when the prison walls collapsed during a lightning strike.

Upon entering the ruins of Frankenstein's former lab both Niemann and Daniel find notebooks but quickly discard them when they realize they are not the records they are looking for. After thawing the frozen Wolf Man Niemann asks Larry Talbot, "Do you know where he kept his records?" After an affirmative answer, Niemann promises, "Show me those records and I will build a new brain for you. I'll lift this curse from you forever."

Talbot helps locate the notebook, which is about 8" × 12" in size and appears to be about 100–125 pages long. The title seen on the cover of the notebook is: "Experiments in Life & Death. Henry Frankenstein." Niemann goes through the notebook looking for ways, "...to combine Frankenstein's technique with mine." This is typical thinking for scientists in that they refer to a colleague's previous notes (most often in the form of a scientific paper published in a science journal) and then combine the techniques and results with their own work to advance and extend their knowledge.

After Niemann successfully refurbishes and re-stocks his lab to do all his brain transplant work we get a good look at a small bookcase abutted against a back wall. The top shelf is loaded with glass filled containers whereas the lower shelves are stacked with many volumes of both books and varying sizes of notebooks clearly showing that Niemann does indeed have his own copious notebooks that he can "combine with Frankenstein's." Later, when talking to Ilonka, Daniel draws a "pentagram" in an open notebook on a lab bench. The pentagram he drew does not look like a pentagram but, rather, more like a hexagon. Inexplicably, moments later we see the notebook page again and the drawing now looks like the front of an A-frame doghouse! Notebooks are neither artist's sketchbooks not scratch paper. And, lastly, over the fireplace mantle in the refurbished Niemann lab are stacked books and notebooks. Apparently, there notebooks are everywhere.

Abbott & Costello Meet Frankenstein (1948)

Insurance investigator Joan Raymond (Jane Randolph) finds "'Secrets of Life & Death,' by Dr. Frankenstein" hidden in a secretary writing table. The notebook is about 8" × 14" and 125–150 pages. Raymond opens the notebook and we can clearly read, "I am sure I can harness the lightning and extract its life giving power. I need but one more part to prepare the monster for my final experiment. Tonight we shall steal another body. All my research is based on the premise that all things, even thought are material." Raymond continues to peruse several more pages of the notebook and those we see are just words with no drawings, figures, images, or even inserted pages. This suggests this notebook is a compilation or summary of some recent events and not the original material of Dr. Henry Frankenstein.

The plural "we" from the notebook commands attention, implying that Dr. Frankenstein had at least one accomplice (Fritz?) in obtaining the bodies he needs for his experiments. Could there have been other helpers or assistants?

Summary

Dr. Henry Frankenstein and Elizabeth were apparently a fertile couple since they had at least two sons, Wolf and Ludwig, plus a daughter, the Baroness Elsa. Based on all the various sizes seen in the Universal Studios' Frankenstein films it appears there were several notebooks originally made by Dr. Henry Frankenstein and each of these seemed to end up being discovered or reused in subsequent film sequels by the sib-

3. *The Notebooks of Frankenstein*

lings. But of all these notebooks which one is the real and original? Even I am not sure of that one. Perhaps it has yet to be discovered and still remains in the ruins of his lab (somewhere on the Universal lot).

From these seven Universal films we can construct a notebook genealogy tree. The original notebook is that of Dr. Henry Frankenstein. During the course of describing and monitoring the early days of his monster, his mentor, Dr. Waldman, also contributed some notes. In the first sequel, *BoF*, Dr. Pretorius also contributed to the original notebooks. After the lab in *BoF* was destroyed Dr. Henry and Elizabeth got busy creating their family. First son, Dr. Wolf (*SoF*) inherited the notebook that he eventually abandoned when he left the village after the monster was pushed into the sulfur pit. Dr. Henry's second son, Dr. Ludwig (*GoF*), inherited the notebook from his brother who used it in his own brain research. After Dr. Ludwig's lab was destroyed the notebook was abandoned and left in the castle ruins. Dr. Henry's daughter, the Baroness Elsa (*FMTWM*), knew where this diary was kept and showed it to Larry Talbot and Dr. Mannering. At the film's conclusion, the castle was washed away by a dynamited dam and supposedly the notebook was too. However, in the next film of the series, *HoF*, Dr. Niemann finds the records (remarkably not waterlogged) and uses them to enhance his own brain transplant work. Since the villagers torched Dr. Niemann's lab we can only assume that maybe the notebooks of Frankenstein were finally destroyed. But since we see a notebook in the *A&CMF* film then someone found it. However, it must be pointed out that these notebooks survived explosions (*BoF*), fire (*GoF*, *HoF*), abandonment (*SoF*), and water (*FMTWM*) so maybe they have more lives than cats and Dracula combined. We may see this notebook again.

4

The Chalk Notes of Dr. Gustav Niemann

In *House of Frankenstein* Dr. Gustav Niemann (Boris Karloff) didn't have access to a traditional notebook of any sort, so he compensated by writing his notes on his Neustadt Prison walls using chalk.

Chalk

Natural chalk is composed of calcium carbonate ($CaCO_3$) and is a soft compact calcite mineral. Natural chalk was made from the tiny skeletons of minute plankton in the ancient seas. After the plankton died their skeletons sank to the bottom and ultimately formed chalk covering vast areas of the ocean floor. Most of this chalk came from the Jurassic Period, about 150 million years ago, so it is very old. These plankton skeletons are primarily those of ancient bivalves, foraminifera, and ostracods, the most dominant being from the genus, *Globigerina*. For those of us who are old enough to have been instructed by teachers using actual blackboards, the chalk they used was from Globigerina.

The Notes

The first photos I saw of the film were in the monster magazine *Famous Monsters of Filmland*. In the background of the stills of the jail cell scene were the chemical symbols that I ignored while first watching the movie. After taking some college courses, I saw the film again on TV and immediately realized that some of the chemical equations were wrong, which prompted me to take a closer look.

In the opening scene of *House of Frankenstein*, we see the prison cell where Dr. Niemann and his associate, Daniel (J. Carrol Naish), are serving their sentence. It appears that even while incarcerated Niemann continues his quest into the scientific unknown, with copious chalk notes scattered on the prison walls.

Over time Niemann needed more chalk to continue writing his notes on his prison walls and he was desperate enough that he sacrificed not only his meal, but

4. The Chalk Notes of Dr. Gustav Niemann

also the enmity of his prison guard by choking him to get more chalk. Getting that chalk overrode his sense of self-preservation and hunger and became his reason for living. Apparently, upsetting a Neustadt Prison guard was just a minor annoyance to Niemann, secondary to his getting that chalk. Which goes to show how important he placed his note taking, way above his concern for his fellow man or, apparently, his own wellbeing (including Daniel's).

After breaking from the choke hold the prison guard's comment of "Try that again and I'll put you in solitary confinement. You would be Frankenstein," seems to have gone unheeded with Niemann suggesting the good doctor has some status at the prison. Not many prisoners can afford such arrogance towards their guards without fear of punishment. My guess is that this was not new behavior by Niemann so he probably behaved like this in the past. After all, he has been in prison for 15 years (see below). Therefore, the guard should have been aware of Niemann's notoriety and behavior and should have taken appropriate measures. Also, if Niemann had a habit of misbehaving towards his guards then this only adds to the confusion as to why the good doctor had such spacious cell accommodations and was not in solitary confinement as the guard threatened.

Opening Scene

The *HoF* film and opening scene of Niemann and Daniel in their Neustadt Prison cell takes around 1:40 minutes of screen time. All in all not much but what is on those walls is! In the reality of the film the only real purpose of all those chemical structures and reactions is to show that the good Dr. Niemann is some sort of genius, above all the mundane necessities of life, and his "he knows what he is talking about so you better pay attention" attitude is supreme and commands your attention. So, what can we learn from taking a closer look at what is really on those walls? Was Niemann a genius or just another wannabe? What we see on those walls is so brief that it may not be worth the effort to take a closer look but we will anyway. After all, this film was made in 1944 and even though many aspects of biomedical science were not well understood at the time the fundamentals of organic chemistry were (organic chemicals are those based on the carbon atom; the majority of our body tissues are made of carbon).

All of this then begs the question as to why did Niemann write his particular notes on his prison walls? What was he trying to save, in typical notebook tradition, on those walls? The walls could be erased so easily. Certainly, a disgruntled prison guard could have punished Niemann and simply washed down the walls thereby effectively erasing all that he had done. However, since what he did write on those walls was so fundamental (and some quite humorous) that he should have no problem duplicating them later in a real notebook because he should have this basic information readily stored away in his brain and, if so, then why go to the trouble to write it on the walls in the first place?

Frankenstein

One possible explanation is Niemann was teaching Daniel some of the science of what he is trying to accomplish. In addition to the basic organic chemistry structures and formulas on the walls Niemann was also teaching Daniel about his brain transplant procedure. Even so, describing the simple organic chemical reactions and structures is not related to the main task at hand, a brain transplant. Also some of these notes (chemical structures) are on the front section of the small internal cell occupied by Daniel making it very difficult for him to effectively see further mystifying their real purpose.

It is noted that Daniel is in what seems like a separate more confined cell within the cell/room occupied by Niemann. This makes me wonder what crime(s) did Daniel commit where he was given even more limited space than Niemann. Was Daniel also convicted of taking bodies from graves?

Later in the film when Niemann and Daniel had captured both of their enemies, Herr Ullman ("my trusted old assistant") and Herr Strauss, Niemann states to Strauss, "...testified he saw me take a body from its grave" and then to Ullman, "...testified for the state." In attempting to bargain his release Ullman states that he saved 15,000 Marks and was willing to give it all to Niemann. Niemann then comments, "...a thousand for every year I spent in a stinking slimy dungeon," so we know that Niemann was in prison for 15 years before his escape. With that being said then there could have been literally years of notes on those prison walls. And no doubt, over the years some of it was erased and replaced with new notes, formulas, and procedures.

Though purely speculation, were all of Niemann's 15 years spent in that same prison cell? If so, then those walls did indeed serve as his notebook with notes and chemical equations a plenty over so many years. How many pieces of chalk had Niemann used over a 15 year period? How many guards were throttled and how many meals were sacrificed in getting his chalk? Also, were any of the notes made by Daniel? He may have been there 15 years too and from time to time was allowed to be in the larger cell and certainly could have made some of those notes himself.

Club Med Prison

Niemann has what appears to be the Club Med of Visaria European prisons. It is interesting that his prison cell is large and spacious enough to have lots of wall space. Most prison cells are in the order of 8' × 10' to about 10' × 15' (by U.S. standards; by Visaria standards, to be sure no health and safety inspectors there, I can only assume that prison cells are far more cramped and crowded). Nevertheless, whatever standing Niemann had at the prison he and his assistant Daniel have quite posh quarters since they were the only two inmates we saw in their large roomy cell. Also, a hanging lit lantern is present during this scene so Niemann must of had special privileges in his prison cell to get such luxuries (who pays for the lantern's oil?).

From analyzing the opening scene in the *HoF* film it appears this prison cell has

4. The Chalk Notes of Dr. Gustav Niemann

at least 11 surfaces that bear notes (ten walls including the alcove of the wood cell door plus the back of the cell door itself). Our first view of some of these notes is when Niemann forcefully takes his piece of chalk from the prison guard. Looking into the cell, through the small cell door window, over Niemann's right shoulder, we get a glimpse of something written/drawn on the back wall; this is the same wall where his straw bed is located. Unfortunately, we do not get any further view of this image so what was actually drawn/written there is unknown. However, the upper blurry image appears to be a head or brain diagram and below it is what appears to be a chemical structure.

On walls in the alcove just to the left and right of the cell door as you enter the prison cell are a series of chemical structures and reactions that have almost no relevance to Niemann's discussions. Some of these notes signify various buffers consisting of simple carbon dioxide and what it does. Also visible are some chemical equations and various organic chemical structures based on carbon (these chemical equations are so fundamental that it is surprising why Niemann wasted valuable cell wall space with such trivial and inconsequential information) and some of these structures are wrong. An example appears on the middle of the left wall where a "CH2"

House of Frankenstein's Dr. Niemann (Boris Karloff) has his chalk. On the walls of his jail cell are many chemical symbols and equations. Some are correct and others quite wrong.

45

should have be a "CH3," the same as the other parts of that structure. On the back of the wood cell door are other chemical equations, some are accurate and some are not, including the chemical equation running down the length of the right side of the wood door from top to bottom, "$CH_3=CH_3OH \rightarrow CH+CH(O_4) \rightarrow$."

After viewing Niemann inside his prison cell at the closed cell door the camera pans right over a wall towards Daniel's cell and as the camera pans right we see some more or less slightly blurry diagrams, including one of a distillation apparatus, and more chemical equations on the wall. The notes on this wall are apparently supposed to signify some sort of organic carbon reactions but they are mostly wrong and do not make any sense. Furthermore, some of the organic chemical structures drawn are wrong and even for 1944 someone should have gotten them right. On the wall below the diagram the chemical equation, "$(R=CO)_2O+2NH_3$" is wrong, though "$(RCO)_2O+H_2O$" does have possibilities.

On this wall, between the door and Daniel's cell, as mentioned above, is a drawing of a glassware distillation setup that has a retort sitting on top of a ringstand to support it and this apparatus is seen over a lit (!) Bunsen burner. A retort is used to distill and concentrate solutions so it is not understood why this would be drawn on valuable wall space since it is irrelevant to the primary work at hand. What sort of solutions would Niemann need to distill and why such elaborate drawings to demonstrate this? Was he trying to teach Daniel some more chemistry? Is drawing a lit Bunsen burner really necessary?

Over the curved wall on the front of Daniel's prison cell are additional chemical structures so the writing was literally on every surface available. The hexagonal shaped structure (benzene) on the upper right wall to Daniel's cell has the six reaction designations of the carbon atoms listed in counter-clockwise order starting at the top where the atom bromine (BR) is attached.

To add to the confusion the size of the various drawings and chemical structures are often unnecessarily large. As such, much wall space was spent on making these large images. If the largeness was so Daniel could better see them (his eyesight may not 20/20) then why so large on surfaces he couldn't see such as the front wall of his cell or the inner walls of the door alcove? If Niemann had made small images then he could of added a lot more notes to those walls.

The centerpiece of the notes has to be the brain/dog/electrical setup diagram. Niemann had drawn a detailed cross-section of a human skull/brain and says to Daniel "this (*human*) brain, taken from a man, and transplanted into the skull of the dog, would give him the mind of a human being." This diagram is reasonably accurate for 1944 and even includes a generator to supply the electricity to jump start the brain. Also stating the obvious, a human brain is too large to fit within the skull of a dog, even the largest breed, so this experiment is doomed from the beginning.

While lecturing Daniel on the planned brain surgery, Niemann commented that Dr. Frankenstein cut the brain stem "here" (which appears to be below the cerebellum area and excluding the spinal cord) whereas he himself would have cut it differently.

4. The Chalk Notes of Dr. Gustav Niemann

(In reality, neither one was right.) When Daniel asked Niemann how he knows this, Niemann responded that he learned what he did from his brother, who, it was explained, was an assistant to Dr. Henry Frankenstein. Could this have been Fritz or perhaps a helper of his? It is noted that the new chalk Niemann earlier wrestled from the prison guard does not match up with previous chalk marks. The "old" wall notes appear to be painted on with a wide brush and not the thinner lines made by chalk.

The eye area of both the man and the dog have been drawn over with what appears to be an eye cover patch similar to those worn while sleeping, the purpose of which is unknown. Most likely this is a function of 1944 sensibility standards, not to reality. The man and the dog would both be anesthetized for the operations so an eye cover would be unnecessary.

The adjacent profile drawing of the human brain in a skull is standard. The anatomical designations of "A," "B," "C," etc., to designate brain compartments are arbitrary and probably a teaching device for Daniel. Niemann probably learned from his brother, via Dr. Henry Frankenstein, that within the brain are certain sub-anatomical areas important for normal brain physiology and he wanted to demonstrate this to Daniel. Nevertheless, the overall brain structure with its haphazard anatomical subdivisions suggests to me that Niemann is not a brain expert.

$H_2O = H_2O$

Just before the prison walls cave in as a result of the electrical storm raging outside Niemann runs to his straw bed and cowers on it as the walls crumble. On the walls over his bed are various chemical and algebra equations. On a wall above the length of the bed are some interesting equations involving the atom bromine; shown as "BR" in the scene but its real chemical abbreviation is "Br." These chemical equations are gibberish. At room temperature elemental Br is a fuming red-brown liquid and is toxic and corrosive. In nature, it is mostly found in a non-toxic salt form, much like sodium and iodine. It is a mystery why bromine was chosen. The most dominant current industrial use of Br is as a fire retardant. In mammals, Br has no essential function. At one time, primarily before World War II, bromides were used as a widely prescribed medical sedative. Currently, bromine is used as an antiepileptic. However, organic molecules with Br atoms attached do have uses in the synthetic drug industry. Also on the wall above his bed are some algebraic equations that do not make sense. On the wall abutting the head area of his straw bed is another organic carbon structure that is wrong.

After seeing some rubble come crashing down in the cell in the above described scene there is an edit cut away from Niemann showing more cell wall crumbling. Then, we see a quick edit cut back to Niemann which is the second shot of him cowering on his bed. Only this time what is written on the corner walls above his bed

Frankenstein

is completely changed. Now, we can clearly see the funniest equation in the entire prison cell scene: "$H_2O=H_2O$"! Just goes to show how brilliant Niemann really is! (Or was this written by Daniel when he was allowed in the larger room in an attempt to show Niemann his chemical smarts?) Obviously, this scene was filmed at a later time and some set dresser just put whatever came to his non-chemical mind on the wall thinking everyone would be looking at Niemann and the falling debris and paying no close attention to the writings on the wall. The other amusing chemical equation is just below the water equation where it is written, $H+Cl=HCl$, which is the chemical abbreviation for hydrochloric acid. Another brilliant piece of chemical insight by the good doctor (or maybe this was Daniel again). If all of this sub-elementary school level chemistry was for Daniel's benefit then no wonder it took 15 years to get him up to the level of understanding that water equals water. Lastly, the notes on the same wall abutting the head area of the straw bed were also changed. This time there is an over simplified (and incorrect) chemical equation using iron atoms (chemical symbol is "Fe").

At the end all of his notetaking was for naught since his cell was destroyed by a lightning strike and the walls caved in. The good news is the cave-in gave him his freedom so he could search for the real records of Dr. Frankenstein. Apparently the notes he made on his prison walls were not significant enough because they were not referred to again throughout the rest of the film. Since his prison wall chalk notes were so simple he had no real need to re-write them again and should have had them readily stored in his mind. Once Niemann did find the notebook of Frankenstein then he had no real need of his own notes since those of the good doctor were superior.

So, with a film made in 1944, with seemingly simplistic chalk lab notes on prison cell walls, we can still enjoy what was written by the set dressers. After all these years this film continues to reveal its many charms. This particular brief opening scene in the prison cell was to set the stage for the rest of the film, which could be subtitled, "a search for the notebook of Dr. Frankenstein, 'Experiments in Life and Death,'" and not be an accurate take of real world science. The major plot of the *HoF* film is the search for the records which included the discovery of the frozen monster and the Wolf Man that subsequently lead Niemann to set up his previous laboratory for the fiery climax. The prison scene was to show that Niemann was indeed a bad man (putting a human brain into a dog's skull classifies him as such) and his escape gave him the freedom and motivation to seek out his enemies and to re-start his scientific career.

5

The Laboratory of Dr. Septimus Pretorius

The rich role that science plays in *Bride of Frankenstein* (1935) has been written about extensively, but the laboratory of the nemesis of Dr. Henry Frankenstein (Colin Clive), Dr. Septimus Pretorius (Ernest Thesiger), and the experiments therein, have been mostly overlooked.

Much has been written about the 1935 film, *The Bride of Frankenstein*, and for good reason. This film is considered the best example of cinema horror there is with the acting, atmosphere, set design, style, and production all top of the line. What has not received as much attention is the laboratory of Dr. Henry Frankenstein's nemesis, Dr. Septimus Pretorius. It is the visit to Pretorius' lab that ignited the scientific spark of curiosity in Henry Frankenstein. Maybe it was the chemical odors, the view of all those homunculi, or how the other elements of his lab were all organized and structured, that got his scientific curiosity juices flowing again. Then again, perhaps Pretorius's comment of "...a woman. That should be really interesting!" was enough to make Frankenstein wonder if maybe he could ... just once more. So, let us take a closer look at Pretorius' lab and understand what he was doing and what he had available to work with. Was what he had enough to do the job at hand?

For Pretorius to do all the work implied in this film he would have to master many areas of biomedial science. The implied scientific disciplines necessary for Pretorius' to achieve his ultimate goal, namely, the creation of a fully functional female brain are anatomy, cell cultures, development biology, physiology, molecular biology, neuroanatomy and neuroscience. Only in a world of film can a cinemascientist be so talented. But, for the moment, let's keep the switch to our Jacob's ladder off as we focus on the immediate issue.

Septimus the Man

We first see Dr. Septimus Pretorius as he appears at Baron Frankenstein's doorstep where he orders the maid, "Tell him Dr. Pretorius is here, on a secret matter of graaave importance!" The maid escorts him to the Baron who further identifies himself as an old university acquaintance, a doctor of philosophy (and in his own words, also a doctor of medicine). In response to an unasked question, Pretorius further explains, "Booted out. Booted, my dear Baron, is the word ... for knowing too much!" Knowing too much certainly has a sinister edge to it and based upon what happens later in the film one can speculate that knowing too much involved his alchemical work with homunculi.

Pretorius plays a renegade mad scientist to the hilt and his personality helped to define what it means to be a cinema mad scientist, especially during the 1930s. He has no morals and even less scruples with no regard for human life or ethics and only seems to care about his own prestige and reputation. Bound up with his on screen persona is an undercurrent of homosexuality as well as a character not far removed from the homunculi devil all indicating that Pretorius and therefore, all that he does, is considered evil.

Pretorius is gleefully played with great panache and flamboyant style by Ernest Thesiger. He is a tall, emaciated-looking man with an unusually large nose and devilishly pointed ears, who embodies the mid–1930s image of the cinematic mad scientist, a role he clearly relishes and doesn't dispute, prodding Frankenstein, "You think I'm mad? Perhaps I am!"

Mad he is, for Pretorius wants Henry to help assemble and bring to life a female equivalent of Frankenstein's monster. Frankenstein is to build the body while he supplies the brain grown "from seed." The fast talking Pretorius tries to convince Frankenstein that they should collaborate, saying "Now think. What a world astounding collaboration we should be. You and I, together! ... Create a race, a man-made race, upon the face of the Earth. Why not?" Pretorius resorts to extreme measures to insure his cooperation, including extortion and kidnapping, but such actions weren't strictly necessary, as revealed by Frankenstein's tormented reply, "I must know!"

Pretorius is dressed as a dark priest, complete with white collar and black skull cap hat. The character is based in part on the historical figure, Philippus Theophrastus Aureolus Bombastus von Hohenheim, better known simply as Paracelsus (1493–1541). Paracelsus was a Swiss alchemist and physician who made a number of significant contributions to science.

Paracelsus also claimed to have created homunculi from human seminal fluid and it is these claims that are at the core of Pretorius's experiments. Pretorius made his homunculi "from seed," perhaps a nod to film censors since using the word, "semen" would have been forbidden. Paracelsus and his work were quite well known and including an allusion to his work brought in a hint of verisimilitude.

Homunculus is a Latin word for "little man" (plural homunculi) and refers to a type of artificial human created through alchemy. Typically, these tiny creatures are grown in jars and are vaguely human shaped. According to the alchemists homunculi could serve as a companion, helper, or possibly as a surrogate child. How the creature was made and the particular skill of the creator determines the type and shape of all homunculi. Also, some claimed to be able to speak but most are not and secondary characteristics such as wings, beaks, claws, and fur are possible.

5. *The Laboratory of Dr. Septimus Pretorius*

Though Pretorius and Frankenstein plot together the real driving motivation and push for the work is the fast talking and extortionist/kidnapper, Pretorius, who convinces Frankenstein that they should collaborate. As Pretorius says, "Now think. What a world astounding collaboration we should be. You and I, together! Leave the charnel house and follow the lead of nature, or of God if you like your Bible stories. Male and female created He them. Be fruitful and multiply. Create a race, a manmade race, upon the face of the Earth. Why not?" Though this did intrigue Frankenstein it wasn't quite enough so Pretorius resorted to extreme measures to insure Dr. Henry's cooperation, including extortion (revealing to the authorities that it was Dr. Henry who was responsible for the existence of the Monster and all his subsequent murders) and kidnapping (Henry's fiance, Elizabeth). In spite of this extra force to get Frankenstein to cooperate Pretorius really did not have that far to go to convince. Such enthusiastic responses as when Frankenstein says, "I must know!" shows that he is already there and esentially onboard with the idea. The passion and curiosity of a scientist is well demonstrated when Frankenstein said this line in the film (and it should also be noted that this curiosity seals his fate).

It is implied that Pretorius was a former teacher of Frankenstein while he was in medical school. During one exchange Pretorius says, "I was hoping that you and I together, no longer as master and pupil, but as fellow scientists...." It is the "no longer master and pupil" line that is the tell-tale remark. Since Frankenstein apparently was a former pupil of Pretorius the "master" sought out his student after learning that the Monster survived the burning windmill from the first film. It is tempting to speculate that Pretorius instilled the "giving life to the dead" experimental mentality in Frankenstein while he was a student under his tutelage. If so, then when during his school training did Frankenstein switch from Pretorius to Waldman? (Waldman was prominent in the original 1931 film, *Frankenstein,* and was also a mentor to Frankenstein.) It is also tempting to speculate that Pretorius was "booted out" of his university position by assisting Frankenstein in his "unholy experiments" of bringing the dead back to life. With this master and pupil relationship then, as a student, how much did Frankenstein contribute to the basic homunculi techniques of Pretorius? Was Frankenstein the pupil a source of inspiration for Pretorius the teacher/experimenter?

While tempting Frankenstein to help him create a "mate" Pretorius wants Henry to build the female body with Pretorius supplying the brain grown "from seed." Quite an interesting and effective collaboration between two scientists who both have something real to bring to the work at hand. Frankenstein would stitch together the parts for a female body, his particular surgical specialty, whereas Pretorius would prepare the brain, his specialty. This then brings up an interesting question of where did Pretorius grow the brain from seed? Did he do this in his lab and then transfer it to Frankenstein's lab or was all the work done in Frankenstein's lab? As Pretorius noted, his lab is not far from Frankenstein's estate when they went there for the first time so the transport of a living brain may not have been that much of an issue.

For Pretorius' opening arguments in trying to convince Frankenstein to join him in his work he says, "I was hoping that you and I together, no longer as master and pupil, but as fellow scientists, might probe the mysteries of life and death ... to reach a goal undreamed of by science ... but you and I have gone too far to stop [what does this mean? 'Gone too far' implies a lot of science was done; who defines 'too far'?]. Nor can it be stopped so easily. I also have continued with my experiments. That is why I am here, tonight. You must see my creation" (note: singular use of word). In response Frankenstein says, "Have you also succeeded in bringing life to the dead?" already showing his interest in what Pretorius has to say. Pretorius replies with, "After 20 years of secret scientific research and countless failures, I also have created life as we say, 'in God's own image.'" With his methods Pretorius has created life in a way that is quite different from both Dr. Henry and God by growing his humans from seeds. And a very effective way of showing how evil Pretorius really is. On the surface he seems to be on par with the work of Dr. Frankenstein in trying to re-animate dead tissue though there is room for interpretation of this.

The "20 years of secret scientific research" should have taught him much. Also, "countless failures" means that in addition to what he did learn he also learned "what not to do" in his research. And if he was actually doing this work for 20 years then he was doing this work while as a mentor to medical student Henry Frankenstein as discussed above. Makes one wonder if Pretorius may have accidentally slipped some key knowledge to young Henry during his student days ("master and pupil"). And the converse may have been true in that young Henry could have told his mentor some information in confidence or perhaps in an off-hand manner that gave Pretorius some much needed insight. All this suggesting to Pretorius that maybe sometime in the future this bright young medical student may be of some use to him.

Also, where was the 20 years of secret scientific research done? Presumably this was done in Pretorius' own lab since it is private and away from those who booted him out for knowing too much. He can do whatever he wanted without anyone scrutinizing his work in his private lab. Also, he could have moved from place to place during this 20 year period, maybe avoiding the authorities or perhaps nosey neighbors.

But Pretorius's work of twenty years, culminating in an impressive collection of homunculi ultimately didn't prove satisfactory. Pretorius dreamed of greater glory: the creation of a fully functional, normal-sized human female. And for that, he needed Frankenstein.

"Visiting My Humble Abode"

We first see the laboratory of Dr. Pretorius as he and Frankenstein begin to formulate their plan. Pretorius gets a bottle of gin and says, "Gin. It's my only weakness" and toasts, "To a new world of gods and monsters" and drinks. Unnoticed by Pretorious, Frankenstein does not take a drink, thereby invalidating the toast.

5. The Laboratory of Dr. Septimus Pretorius

It appears the main focus of Pretorius' lab work was the creation of his homunculi since no other useful items were seen or discussed. Therefore, was his lab layout satisfactory for the job at hand, namely homunculi creation? According to the alchemist, Paracelcus, sperm was needed as a necessary ingredient in the creation of his homunculi so, no doubt, Pretorius used a similar procedure (after all, his work came "from seed"). Mixed in with the sperm seed were other ingredients, many of which were derived from distillates of various plants such as the mandrake root (a herb that has been intimately connected with humans and fertility since pre-historic times), a common component according to homunculi lore. All of these "procedures" were part and parcel of the alchemist's materials and methods. To make such distillates various types of glassware are needed. What is seen in Pretorius' lab, mostly exotic looking glassware, large retorts, distillation columns, and other flasks and beakers, may be just the right stuff for him to carry out his "secret scientific research."

Glassware is primarily used for work on small molecules in liquid form and the ingredients Pretorius would use for his homunculi would be a mixture of such small molecules. All in all it does make sense so based on what is shown in this scene the visible glassware does serve a function for the work at hand and is useful for making the necessary reagents for homunculi creation, such as buffers, various solutions, protein extracts, and maybe "germ plasma," "biological seed" (ie, DNA) extracts. However, since Pretorius is working with living animals (albeit, miniature humans) then he should also have vivarium capabilities, meaning some sort of functional housing (food, water, waste removal) to care for his homunculi. Presumably, this was all kept in a back room, maybe the same room where the homunculi box is kept.

Stating the obvious, from Pretorius' lab we know what he has because we can see it. Visible are a variety of glassware, glass separation columns, shelves of books, work tables, reagents, apparatus, test tubes (with cotton plugs) and racks, rubber and copper tubes, heater, trunks, many flasks (erhlenmeyers and round bottom), an alcohol lamp (no Bunsen burner in sight), and a sink. As well as the proverbial human skeleton. All essential items for just about any lab and very typical for a mid–1930s cinema lab. In many contemporary films much of the bench bling is just that, bench items that look cool as eye-candy and interesting but have no real bearing on the work at hand. Not so with this film since what we see in Pretorius' lab is very much in keeping with his presumed work.

A relatively small monocular microscope is briefly seen as Pretorius is describing his ballerina homunculus. The microscope, seen on a back bench and is slightly out of focus in the scene, is visible over Preotrius' right shoulder. This microscope would have been inadequate for the initial work of growing cultures "from seed" since Pretorius would have needed more sophisticated optics with larger magnifications than what this small one could provide.

Retorts are used to help distill various types of liquids and since these unique looking glass vessels are used in the various alchemical prodedures to create homunculi we see several of these in Pretorius' lab. The largest one and the most prominent

Frankenstein

one on the film set, on a separate table, looks like it can accommodate all the fluids necessary for the homunculi that he created. This giant retort is used for distilling complex liquids to get evaporates (to make homunculi?), typically for distilling and separating small molecules; the large ehrlenmeyers are hooked up to the retort to receive the distillates. In the creation of homunculi there would be a lot of mixing of liquids and fluids. To make some of these fluids, such as salt buffers, a mortar and pestle (rare now) would be used to grind the large salt crystals into smaller pieces to easier dissolve. A thistle tube is attached to the top of the retort that is used to add additional solutions and fluids to the mix.

Throughout the lab are cast deep Caligari expressionist-like shadows. In the very back of the elongated lab is a small bed. On a back ceiling wood beam just above the

The laboratory of Dr. Pretorius (Ernest Thesiger, right) *Bride of Frankenstein.* **Henry Frankenstein (Colin Clive, left) has arrived to discuss creating a bride for the Monster. In the background are interesting "tools of the trade" that Pretorius has used in his "20 years of secret scientific research." The most visible piece of equipment is the large retort flask on Pretorius's table. This retort evaporation-distillation setup has a small, low-heat alcohol lamp warming the retort causing a slow, methodical evaporation. This suggests that Pretorius is separating small and similarly related organic molecules, perhaps some ingredient or nutrient for his homunculi. On top of the retort is a thistle tube that is used to add various liquids directly into the retort without opening the top. Behind Henry is something every respectful mad scientist should have in his lab, a human skeleton.**

5. The Laboratory of Dr. Septimus Pretorius

bed is what appears to be a mold of a human hand seemingly tacked to the wood. Also, on the wall opposite his bed all by itself is a head form (plastercast?). All in all, very modest accommodations.

"My Trifling Experiments"

After his toast to new gods and monsters Pretorius proceeds to a back room to show Frankenstein the results of his work, saying, "The creation of life is enthralling. Distinctly enthralling, is it not? I cannot account precisely for all that I will show you … but perhaps you can." He then goes into a back room and brings out a large oblong wood box that house the homunculi. As a scientist I am uncomfortable with the comment of not being able to "account precisely" for his results. This is a troubling comment since any good scientist goes out of his way to account precisely for all that he does. Without knowing a precise mechanism then all his results can be considered phenomena observations without knowing or understanding what he did. In the real science world after a phenomena has been discovered or observed then the how and why will quickly follow. Giving Pretorius the benefit of a doubt one interpretation is he was stuck on a particular scientific problem and sought the advice of a colleague to provide some insight or direction for future work ("perhaps you can"). On the other hand, he could have gotten very lucky and has no clue how he created his homunculi and thinking that maybe Frankenstein could help him figure it out.

Pretorius keeps his homunculi in a coffin-like box with no apparent light source. As he removes the lid some telltale tubes, glass items, and small machinery, probably pumps, can be seen. On one side of the box, there are indentations of the seven homunculi jar lids on the underside interior velvet lining. The other side of the box contains other jars, dials, and connecting tubes that presumably serve as life-support systems providing oxygen, food, and water, as well as sanitation removal. These would be important to maintain not only oxygen levels and eliminate waste products, but also maintain nutrient levels in each of the homunculi jars. The ecology and necessary support for the mermaid would certainly be different from that of the others due to the aquatic environment.

The Seven Homunculi

Pretorius created seven different homunculi. While removing the jars Pretorius says, "My experiments did not turn out quite like yours, Henry. But science, like love, has her little surprises, as you shall see … there is a pleasing variety about my experiments. My first experiment was so lovely that we made her a queen."

Frankenstein

That one word, "we" implies that Pretorius had helpers. Were these helpers technicians or fellow scientists? Or perhaps someone far more sinister and unscrupulous like Pretorius (such as a partner in crime whom Pretorius may have unceremoniously dispatched for knowing too much). Also, his comment about his experiments not turning out quite like Frankenstein's implies that Pretorius failed at making full sized humans. So, why were his creations so small (they all seem about $\frac{1}{12}$ to $\frac{1}{8}$ scale)?

- Queen: a label on the jar reads: EXP λ GA178; on command she performs a mechanical windup curtsy.
- King: a label on the jar reads: EXP Y VXY7; he seems to be patterned after Charles Laughton's Henry VIII.
- Archbishop: a label on the jar reads: EXP λ PU8; he blows a whistle and rings a bell at the King's amorous antics towards the Queen.
- Devil: a label on the jar states: EXP λ 1403. Pretorius inquires, "There's a certain resemblance to me, don't you think? Or do I flatter myself?"
- Ballerina: the jar label is not visible. Pretorius comments that her talent "is charming, but such a bore."
- Mermaid: no jar label. The mermaid is seen comfortably resting under water on a rock. She is combing her long hair with a seashell comb with undersea light dancing on her sequined tail. Pretorius explains, "You can never tell how these things will turn out. It was an experiment with seaweed."

Though not shown from the front, a parting shot reveals a seventh homunculi jar with a child (adult actor Billy Barty) on a high chair waving at the good doctors.

It's the last we'll see of Pretorius's homunculi as he walks towards Frankenstein, elaborating on the fundamental limitation of his experiment, "Normal size has been my difficulty. You did achieve size (obvious since Frankenstein started with full size body parts). I need to work that out with you. You did achieve results that I have missed."

What is interesting about this comment is that Pretorius is freely admitting a shortcoming and a flaw in his own "20 years of secret scientific research" by saying in essence that "size does matter" and he does not know how to do that. He has had 20 years of "countless failures" to learn from but apparently not enough to actually succeed. Again, Pretorius demonstrates that he is clearly in charge and telling Frankenstein the methods he is missing are those that his partner will contribute one way or another.

After seeing the homunculi an astonished Frankenstein says, "this isn't science, it's more like black magic" and this statement deserves comment since there is more to it than meets the eye. This comment from Dr. Henry suggests, within the context of the film, that he does not think outside of the box. Dr. Henry does not understand the biology necessary to create homunculi so to him it is indeed black magic. (I am reminded of the famous "third law" quote from Sir Arthur C. Clarke, the science fic-

5. The Laboratory of Dr. Septimus Pretorius

tion writer, who said, "Any sufficiently advanced technology is indistinguishable from magic." The advanced science technology of Pretorius seems as magic to Frankenstein and he reacted as such.) All in all quite out of character for Frankenstein since he can be considered the King of out-of-the-box thinking.

In creating his homunculi it should be pointed out that Pretorius also made miniature clothing (impressive bead work at that scale, not to mention the miniature jewelry!) as well as a miniature turkey leg for the King to munch on while pursuing the Queen. In addition, a miniature whistle and hand bell were made for the Bishop and miniature rock formations, not to mention that seashell comb, made for the mermaid. Perhaps Frankenstein was commenting on the homunculi couture and attire ("black magic") instead of the creatures themselves.

"Grew Them as Nature Does ... from Seed"

In one exchange with Frankenstein, Pretorius says, "While you were digging in your graves piecing together dead tissues, I, my dear pupil, went for my material to the source of life. I grew my creatures like cultures, grew them as nature does ... from seed. Leave the charnel house and follow the lead of nature." Nature, of course, sets the gold standard of scientific accomplishment. It is wise to try not to improve on Nature but, rather, try to equal and replicate the exquisite design and function that has taken countless millennia to optimize.

Pretorius' revelation, "I grew my creatures like cultures, grew them as nature does ... from seed" warrants some scientific inquiry. What exactly does he mean that he grew the creatures like cultures? Initially, his homunculi would all start out as cultures, but perhaps not what you may imagine. A proper explanation requires a little background in embryo and tissue development, the events that occur after an egg has been fertilized by a sperm cell.

Body growth is based on patterns and this includes organ development such as the brain. Nature is full of ordered patterns with some structures more complex than others. Higher organisms such as mammals, including man, have patterns of increasing complexity. All of this coordinated growth and development is genetically determined so bio-design, including brain development, is controlled by one's own DNA. The idea that "It's all in the genes" is very apt here.

After fertilization, when mammalian sperm and egg cells combine ("grew them as Nature does"), the process of embryo growth or embryogenesis, quickly begins. The first few cells organize themselves into patterns of tissue architecture and as the embryo grows these patterns are of increasing complexity. Continued growth occurs as self-organization of cells and tissues into organs, including the brain, begins. Self-organization is the process of spontaneous formation of ordered patterns and structures from elements that have no or minimal patterns, meaning different cell types non-randomly come together. Also, during development many of the cell types can radically change or morph into a completely different cell type. All are normal phases of normal growth.

Tissue self-organization can be classified into three categories: self-assembly (time evolving control of relative cell positions, such as the layered pattern in tissue development), self-patterning (spatiotemporal control of cell status, so that cells acquire heterogeneous properties from a homogeneous cell population like all the different types of brain cells that result from simple precursors), and self morphogenesis (local growth and remodeling such as seen with eye formation and some tissue mechanics); these processes are not necessarily independent of each other. Self-driven morphogenesis involves complex controls of tissue stiffness and viscosity. Local interactions are complex in tissue self-organization. Therefore, a change in a cell's state (such as in differentiation when one cell morphs into a different cell type) could simultaneously cause an alteration in the interacting rules meaning cellular development is not static but quite dynamic. Tissue scale, cell scale, and intracellular scale all play a role in brain development. All of these could be considered as a cyto-ecology of the developing brain, meaning this environment helps to determine how the cells are hooked together. One problem Pretorius had to master was the ability to steer a multicellular population with complex behaviors into a dynamic, functional brain. His "grow from seed" comment is complex indeed.

During a natural life cycle the combination of sperm and egg, as mentioned above, result in an embryo that eventually develops into a fetus which subsequently comes full term upon delivery. This development is a continuous process and, this is the most important part, this process continues throughout life *after* birth. When a sperm cell combines with an egg cell this fertilized ovum is referred to as a zygote. It is this zygote that undergoes rapid development and growth. At this time certain cells become specialized such as the spine, internal organs, bones, and, of course, the brain. At this time the fetal brain rapidly grows both in size and in the number of neurons it has. It does indeed take some time for all the neurons to properly connect to each other.

Also, for proper embryogenesis to occur there must be some sort of fertilization process. By fertilization is meant the fertilization of an egg and sperm cell resulting in an embryo. With his homunculi Pretorius was able to bypass the fertilization step and able to directly stimulate his "seed" to undergo embryo and zygote growth. For humans the first eight weeks are the most important since that is when all the early organ formation and specialization occurs. After the eight week mark, the growth of the fetus adds more size and substance to the body.

The brain starts out with a small cluster of cells that rapidly, over a period of several months, develops into a functional brain with all the specialized cells. A newborn brain, one that has undergone nine months gestation, is still developing for many years yet to come. In this respect, a developing brain should be considered a work in progress. And this certainly applies to our Bride. Even newly born from electricity, that brain, grown from seed, still needs significant development, which could

5. *The Laboratory of Dr. Septimus Pretorius*

take years. The ability to respond to external stimuli through the senses of sight, touch, hearing, taste, and smell, are learned experiences and something the newly born Bride would not have developed in so short a time span.

Brain Engineering

What would it actually take for Pretorius to grown a fully functional human brain from seed? Even though the creation of homunculi, as originally conceived by the alchemists, is pure fantasy we can nevertheless learn some interesting things from this exercise. Using homunculi procedures Pretorius would most likely start with human semen or "seed" (whose? Pretorius himself or some other male?) and mix it with a variety of substances one of which would be blood (whose? Male or female?). It would be optimal to keep this volatile mixture at body temperature, but a slightly higher temperature would be tolerable, coming from an alcohol lamp or perhaps a Bunsen burner. The fluid would then be incubated, fed and nurtured along during growth.

During brain development the nervous system is derived from the ectoderm cell layer, the outermost tissue layer, of the developing embryo. During the third week of development distinct patterns begin to form and one of them, the beginning nerve tissues, called the neuroectoderm, begins to appear and eventually forms the neural plate which becomes the neural tube, the source of the majority of nerve cells in an adult. Our brain and spinal cord come from this development. At this delicate stage any mutations and/or injuries that occur can lead to debilitating or lethal deformities. The closest part of the neural tube to the forming head, above the neck region, eventually becomes the brain. Newly formed and developing brain cells then further separate into neurons and glial cells, the main cellular components of the brain, as well as other anatomical regions such as the pons, hindbrain, and the various hormone-producing organs like the hypothalamus. These newly formed neurons move other parts of the developing brain to self-organize into the different brain structures all controlled by axons, dendrites, and nerve synapses. It is the connection and communication between these nerve synapses that control sensory and motor movements, all behavioral elements. In order for his homunculi experiments to succeed, all of these delicate and elaborate procedures had to be carefully orchestrated, monitored and controlled by Pretorius via his alchemical methods.

As for the brain of the bride, how did Pretorius know the brain he grew from seed was a female brain and not a male brain? After all, using sperm seed cells as an ingredient in his procedure would suggest that both X (female) and Y (male) chromosome bearing sperm were present, so there was more or less an equal chance of a male or female brain resulting. Also, how many attempts did he make to grow the brain used? Did he start several seed cultures and took the one that looked the most promising?

Frankenstein

Neural Development

For the developing brain many of the cell types change or morph into distinct cells with special functions. There are many types of brain cells and the challenge is to control the diffusible molecules, control circuits, and gene regulatory networks so cells know where they are and what their function is. Biomolecules called nerve growth factor and others like morphogens, which are diffusible molecules, help in controlling developmental pattern formation, a key aspect of brain development, function, and physiology. These morphogens and growth factors help steer different brain cells into their precise location much like a tugboat brings a larger ship to port. In this respect Pretorius had to do some cellular engineering in being able to get each cell located to its proper location. Brain cells like neurons, glial cells, Schwann cells, axons, and a myriad of others are all part of human brains and each must be in its proper location for proper function.

Also, Pretorius needed to address the issue of quality vs. quantity. How many brain cells and how to interconnect them all? About 90 percent of the brain cells are glial cells that act as a sort of support for the other cells. Neurons send signals to and from to coordinate muscle movement and there are about 100 billion neurons in a brain (and many, many more glial cells!). This is important in determining the relative IQ of the brain. For Pretorius he can influence the brain's IQ by making sure enough morphogens can move enough cells into their proper location and make sure they are precisely interconnected. Many of these growth signals will accumulate over time. Furthermore, some brain cells are larger than others whereas some are more in numbers than others. In developing a human brain one wonders if certain errors can be fixed before they happen? Can they be fixed "mid-stream"? How about post-maturation when the brain is transplanted into the Bride's cranium? For neurobiologists an important question is whether any brain cells can self-organize or do they need the help of morphogens? Can these cells be spatially self-organized? In culture can Pretorius determine how many cells migrate? Can he determine any cell positional sensing while the brain is growing? Complex indeed and Pretorius had to control all of this.

With all that being said, let us take a little closer look at what processes are needed to generate and shape a functioning brain and nervous system. Could Pretorius control any of the processes of brain development? Could he have increased the number of neurons to make her a "smarter brain"? Please keep in mind that not only did Pretorius have to grow a fully functional human brain but this brain also had to completely integrate itself with the nervous system of the Bride's sewn together body so all movements and processes are controlled and regulated. How much of the Bride's brain did Pretorius grow? How much of the hindbrain, pons, spinal column did he grow? The cellular basis of brain development requires a knowledge of neuroscience and developmental biology and Pretorius would definitely have to "know too much" to accomplish all this. Pretorius was ahead of his time in his ability to understand the

5. The Laboratory of Dr. Septimus Pretorius

cellular and molecular mechanisms to form a complex nervous system. Defects in neural development can lead to cognitive, motor, and intellectual disabilities and Pretorius seems to have all these under control.

Pretorius climbed a much steeper mountain in growing a brain from seed than his partner, Dr. Henry, did in "piecing together dead tissues." A proper brain (normal human brains continue to develop throughout the first 20 years of life) would be constantly developing neurons and precursor cells, migration of immature cells to their final positions, outgrowth of axons and dendrites from neurons, guidance of all the parts to their correct postsynaptic partners, and finally forming all those synapses that give rise to learning and memory.

These neurodevelopmental processes are of two types, active-independent and active-dependent. Active-independent processes are hardwired and genetically programmed by your DNA. Active-dependent processes are those of neural activity and sensory related experiences. Various environmental conditions can affect these processes meaning they are learned over time and not instinctively there. The Bride would therefore not have developed any active-dependent processes.

Along with this brain development is eye development that must also be coordinated. Within the developing brain is the optical vesicle that eventually becomes the optic nerve, retina, and iris of the eye. These nerve connections must be precisely controlled and placed for proper function. Were the Bride's eyes also grown like seed, were they transplanted from a "donor" (another "police case," says Fritz), or came with the original head?

Mixed within all these cell types are glial cells that help guide an estimated 90 percent of migrating neurons to their destinations. In very simple terms, during brain development there is significant cell migration with everything becoming interconnected in its proper way. As you can also imagine, it takes very little to offset this brain development hemostasis into a potentially severe neurological problem. So, Pretorius had to use extreme care in growing his brain. Easier said than done.

Speaking of synapses, nerve cells communicate through small transmitter molecules called neurotransmitters, which are endogenous chemical signals transmitted from a neuron to its target cell, typically a muscle cell, across a synapse. One particular neurotransmitter, called acetylcholine, is a key component of the neuromuscular synapse response. A nerve impulse sends a signal down its length and releases small vesicles containing acetylcholine thereby stimulating a muscular response. These are the details of a synapse and a key element of proper brain function. All of the synapses would have to work in unison for the Bride to properly function.

The size of brain that Pretorius grew from seed is of interest. How was Pretorius able to "super enhance grow" brain development? Furthermore, did his brain snugly fit into the cranium of the Bride? If it was a loose fit then her brain would have jostled around potentially causing some severe discomfort. Also, if it was a tight fit then potential brain trauma and headaches could occur. All in all, there is not much wiggle room for error. We should also keep in mind that any toxic exposure any time during brain development may not only be eventually debilitating but may result in birth defects, even death. These could be considered congenital malformations. Furthermore, a male brain is slightly larger than a female brain so a gender-specific snug fit is preferred.

Not only did Pretorius have to grow a fully functional human brain but it had to completely integrate itself with the nervous system of the Bride's manufactured body.

Along the same line of questioning, did Pretorius implant the grown brain into a detached head which was then attached to the Bride's body, or was the brain implanted into her very head? The few visible sutures seen on the Bride's neck and jawline suggest it was a brain transplant and not an entire head transplant, which would have added complications.

Second View of Pretorius' Lab

As Dr. Henry enters Pretorius' lab the second time we get a good view of the experimental work Pretorius is doing on the table he is seated at. Heated by a simple alcohol burning lamp is a large retort that is connected via a rubber tube to a large collecting ehrlenmeyer flask. The alcohol lamp provides a low though steady heat that would be necessary for a slow, careful distillation procedure. The flask is receiving the distillate from the retort suggesting Pretorius is trying to isolate some small soluble organic substance. Immediately in front of Pretorius is a mortar and pestle suggesting he has been grinding some salts in the preparation of a buffer. Also on this desk are other ehrlenmeyer flasks and other types of glassware, all indicative of isolating and preparing small organic molecules, perhaps something he needs to keep the brain alive or for his homunculi.

Birth of the Bride

During the last minute preparations for the soon-to-be-alive Bride, Pretorius comments, "All the necessary preparations are made. My part in the experiment is complete. I have created by my method a perfect human brain. Already living but dormant. Everything is now ready for you and me to begin our supreme collaboration." Then Pretorius goes on to say, "Lying here within this skull is an artificially developed human brain. Each cell, each convolution, ready, waiting, for life to come." Yes, but how functional will the brain be?

Life's experiences imprint thoughts, concepts, ideas, and, most important, memories into the brain. These are the active-dependent processes mentioned above.

5. The Laboratory of Dr. Septimus Pretorius

Since the brain was newly minted, there were no life experiences or memories. At this point all the Bride had were the active-independent processes that are hardwired and genetically programmed by her DNA.

After Pretorius removes the bandages from the Bride's eyes, we see her moist eyes focus and appear to take in her surroundings. The Bride first moves her right hand. Then she lifts both arms straight out, her eyes close, and her head lowers and rests. This is followed by shaky, jerky head movements, side-to-side, then up, perhaps identifying a light source as the monster did in the original *Frankenstein*.

Her eyes are straight and focused as she takes her first furtive steps and looks around. When she first sees the monster she tries to scream but just utters a choking sound, perhaps a vocal cord problem. Finally, after some effort she lets out a good scream. She looks away from the monster toward Dr. Frankenstein, then she walks quickly to a bench to sit down. The active neuromuscular response followed by an appropriate reaction suggests the procedure has been a success, as this would require a significant amount of mind-body-eye coordination paired with a high level of cognitive function.

Tabula Rasa

Tabula rasa means blank slate in Latin (more accurately, "scraped tablet") which refers to the writing of Romans on a tabula or wax tablet, which was subsequently heated and then smoothed (or "blanked") for new writing. Perhaps an early version of an Etch-A-Sketch. Tabula rasa, in psychiatric terms, suggests that people are born without knowledge, which comes from living life, a balance of experience and perception. In this interpretation nurture is favored over nature when it comes to personality, behavior, and intelligence. So, the bride was "born" a tabula rasa, devoid of learned personality, behavior, and intelligence. She had the nature but did not have any nurture.

Her reactions may not have been in response to her first sight of the monster, but to a fundamental and hard-wired visceral reaction that any visual stimuli might have provided. Were her eyes and brain completely integrated into her spinal column and therefore the rest of her neuromuscular system? It takes time to properly develop control of body coordination. What about breathing? Endocrine issues? (Was the Bride a diabetic?) She screams or rather hisses as she was trying to talk. Her tongue could have got stuck in her throat, or an asthmatic response, or perhaps a vocal cord problem. Her immediate attempts to walk must also be considered. She becomes petulant and refuses assistance in walking. She seems to be ambling, stumbling and falling forward more than walking, suggesting that her neuromuscular coordination is not yet fully developed. A misfiring of the nerve impulses or synapses to the muscles would account for her lack of coordination, similar to muscular dystrophy in that the nerves are no longer properly coordinating muscle movement. Interactions of cells evolve over time and space and couldn't have been 100 percent functional at the time of her "electrical birth" for proper coordinated movement. Contextual, environmental, and interdependent coordination between cells is critical for proper structure and function and the newly born tabula rasa Bride did not have enough time to coordinate all this.

The fate of Pretorius and his remarkable creation, a Bride for Frankenstein, are unknown. In the end, when the Monster decides that both he and Pretorius "belong dead" an explosion occurs blowing the lab to smithereens. Henry and Elizabeth do escape ("You live!" proclaims the Monster on their behalf) and we know that the Monster survived for another sequel (*Son of Frankenstein*). Did Pretorius too escape to begin anew somewhere else under a new guise (à la Neumann/Edleman from *House of Frankenstein*)? It is reasonable to conclude that the Bride was destroyed, but given the penchant for the Frankenstein franchise, if she survived it could be a tale for another time, perhaps, "*The Daughter of the Bride of Frankenstein*," that could be, as Pretorius would say, "really interesting."

Summary

The laboratory of Dr. Pretorius is an important element of the film *BoF* since that is where the Bride's tabula rasa brain was presumably grown from seed. Pretorius' contribution to the Frankenstein mythos was in the creation of a (presumably) functional (female) brain whereas Frankenstein contributed his considerable surgical skills by stitching together her body. All of this work was accomplished because necessity is the mother of invention. The necessity motivation was two-fold: extortion (kidnapping Frankenstein's wife) and the search for science. In their collaboration, Dr. Henry is "annoyed" whereas Pretorius is genuinely "mad" (anyone who uses extortion and kidnapping to achieve his ends should be considered mad). All of the homunculi apparatus and procedures that Pretorius had previously used in his lab helped to contribute to his creation of the Bride's brain and much of this was visible in his lab. The supporting glassware and overall layout of the lab are quite fitting for the work at hand, namely making different reagents, philters, and solutions for the creation of his homunculi. The homunculi scene essentially proves two things. First, that Pretorius does indeed know his science, as corrupted as it is, and second, he has the know how and desire to complete his project, namely, the creation of a fully functional female mate for the Monster. It is assumed that he grew, from seed, the fully functional female brain for the Bride in his lab. Though the glassware seen would have been necessary for the early phase of his work he would need further sophisticated life-support machinery for the development of an intact, functional human brain (such as perhaps seen in the film, *Donovan's Brain*). Lastly, the newly created brain would have to be carefully transported from Pretorius' lab to Frankenstein's lab (don't let Fritz do it since he may drop it...) and transplanted into the Bride's cranium with all

5. The Laboratory of Dr. Septimus Pretorius

ten brain nerves properly interconnected and functional. The newly born Bride was still getting her bearings of sight, sound, and movement, when the lab was blown to atoms. Therefore, with the Bride destroyed we will never know how effective the brain transplant was since further development and experience were needed to see how effective her grown-from-seed brain really was.

6

The Spark of Life
The Popular Science of Mrs. Mary Shelley

"Bring life to the dead."—Dr. Waldman

Mrs. Mary Shelley wrote the book, *Frankenstein*, which was published in the year 1818. Many claim this book to be the first true work of science fiction. Soon we will be celebrating the 200th anniversary of the creation of Frankenstein so we are fortunate to have such a rich two-century history of this iconic story. And much of this history we tend to take for granted which is why it would be of interest to examine the beginnings of the Frankenstein legend to better understand how it all came to pass. Essentially, what was the inspiration that "sparked" the ideas of Frankenstein in a young Mary Wollstonecraft Godwin (nee Mary Shelley) during the early 19th century when she wrote her book?

Mention the name Frankenstein, and all sorts of images come to mind, both of the monster himself and his scientist creator. Images of Frankenstein are global and can be found on virtually anything and everything. For Baby Boomers the visage that most readily surfaces is the face of the legendary Boris Karloff as Frankenstein's monster in the 1931 Universal Studios film. For the scientist, Doctor Frankenstein, there are two competing images. One of Colin Clive (Henry Frankenstein in the 1931 Universal film) and the other of Peter Cushing (Victor Frankenstein in the 1957 Hammer version, *The Curse of Frankenstein*). Both Clive and Cushing defined the role. In the original novel (as well as the Hammer film) the name of the scientist is Victor Frankenstein, which was changed to Henry Frankenstein for the Universal films. However, before all this came to pass it all started with the printed word and parlor room chit-chat.

Popular Science

The popular science of the early 21st century is on DNA and its various mutations. Get the right gene or fix the problem gene and all is good seems to be the pop-

6. The Spark of Life

ular science mantra of today. On the other side of the coin are those whose mantra is "hands off my DNA" so no matter your position it is all about DNA here in the early 21st century. About 100 years ago, during the first part of the 20th century, the new science of endocrinology or hormones had begun so the popular science of the day involved glands and gland extracts. At that time it was glands that could "restore youth" and "bring vigor back to life." And 200 years ago, during the early 19th century, the new areas of electricity, anatomy, and physiology dominated parlor room and pub discussions. And this is where we begin our story about Mrs. Shelley and the popular science of her day and how that influenced her creation. With that in mind, how would it go if her story were written 100 years later during the early 20th century when the popular science of the day was glands? Furthermore, how would the story go now, 200 years later, in the popular science DNAge? Then what did inspire Mrs. Shelley to write her masterpiece?

In the early 19th century it is the presence and prevalence of books as well as parlor discussions that reflected society at the time. So, what books were available to Mary Shelley and what were most parlor room discussions about? It is also important to know what science, during the early 19th century, is known to scientists and of this what science is known by the public and in particular Mrs. Shelley. After all, her book, *Frankenstein*, is a work of science fiction so the science available to her is also important in understanding the background that led to the creation of the book.

Mary Shelley began writing Frankenstein mid–June 1816, so what popular science was known during the early 19th century is what she knew and what she wrote about. When Mrs. Shelley unleashed *Frankenstein* onto an unsuspecting world with the hidden message of "some things man was meant to leave alone" she brought together the popular science of her day which were alchemy, anatomy, galvanism, and electricity mixed with some body snatching. In her day these topics were new concepts and as such these topics dominated many discussions as pop culture of her time.

This has parallels in today's world where the "spark of life," our genes and DNA, today's popular science, mixed with genetic manipulation mirror the early 19th century's concepts of doing things man was meant to leave alone. After all, 200 years after Mrs. Shelley, here in the early 21st century, we have the power to play God by making genes, adding genes, mutating genes, and/or subtracting genes. Scientists can now make species faster than God and how blasphemous is that? According to some these are certainly things man was meant to leave alone. However, genetic manipulation ultimately means better health so in this respect we are meant to manipulate our genes to improve our ability to thrive.

Mary Shelley's Early Education

During the 18th century a period called "The Enlightenment" or "The Age of Reason" occurred in Europe, a cultural revolution in which there was a separation between the traditions of the Church and science, and free thinking was encouraged. William Godwin, Mary's famous father (a writer and political activist), hosted many gatherings at his house with the intelligentsia of the day, including scientists and philosophers, such as Samuel Taylor Coleridge, Thomas Paine, Humphry Davy, William Wadsworth, and poets Lord Byron and Percy Bysshe Shelley. Fashionable topics included the concepts of animal electricity, galvanism, its role in physiology, and the possibility of using it to revive the dead. Other conversations revolved around the human soul and how it might be revived after death. Many thought the heart was where the soul was located. The work of Humphry Davy showed that the function of the heart was to pump blood, displacing the now adrift soul.

Also available to a young Mary were the contemporary books in her father's library. The conversations and books made an impression on young Mary, for a scant few years later at Villa Diodati on the shores of Lake Geneva in Switzerland she penned *Frankenstein,* incorporating many of the concepts she overheard and read in her father's parlor.

The Creation

In Mrs. Shelley's book the actual creation of the monster was established in a single sentence. From the 1818 first edition, "With an anxiety that almost amounted to agony, I collected the instruments of life around me, that I might infuse a spark of being into the lifeless thing that lay at my feet" (Volume I, Chapter IV, first paragraph of 1818 edition).

Since the monster was birthed by a "spark" strongly suggests an electrical theme and therefore implies she knew about the popular science of electricity and galvanism of her time. The new discoveries of anatomy, especially the indication that the heart pumps blood and is not the seat of the soul, brought new dimension in man's relation to nature and God that Mrs. Shelley incorporated into her book. Lastly, also implied in the work are elements of alchemy ("the instruments of life").

Electricity

The history of electricity goes back at least to the Ancient Greeks who observed that when fur is rubbed against amber there was a mutual attraction between the two. By the 1600s the electrostatic generator had been invented, the difference between positive and negative currents was noted, as well as classifying materials as conductors or insulators. Also, during this time the word, "electric" was coined from the Greek word, electron.

Contrary to popular belief Benjamin Franklin (1706–1790) did not discover electricity. However, during the 1740s, using ordinary household items, he developed a

6. The Spark of Life

theory of positive and negative charges for electricity. Famously, he also showed that lightning was an electrical phenomenon. Moreover, he invented the lightning rod and described the concept of an electric battery.

The discovery of electrical current depended upon the development of the battery. In 1800 Alessandro Volta (1745–1827) created the first voltaic pile of alternating zinc and copper plates (an early form of the battery) that produced a steady electric current. Volta also showed that linking a positive electric connector with a negative electric connector could drive an electric charge (or voltage) through them and therefore transmit electricity. The word, "volt" is derived from Volta's last name. At the time, experiments with electricity were considered "parlor games" since they were used in demonstrations to entertain without concern for practical application. Such demonstrations became popular in European society during the early 19th century.

Science became popularized during the 18th century due to the explosive growth of knowledge and interest providing lay people an opportunity to become "citizen scientists." This was the first time real science developed a relationship with the public. A popular science practice by citizen scientists during the late 18th century was the "electric theater" in which people could use Leyden jars to discharge electricity and experience an electrical thrill via kissing or perhaps via a human hand holding chain. (These electric theaters may be considered the original Exploratoriums that are popular now.)

Popular science was presented at aristocratic gatherings so all could share the new knowledge. At the time electricity was a wonder and all were amazed at the tricks that could be performed with it. And Mary Shelley participated in these experiments not only at her father's home but also with her contemporaries like Lord Byron and, of course, Percy Shelley.

Galvanism

Luigi Galvani (1737–1798) discovered bioelectricity, which he named, "animal electricity," by showing that nerve cells use electricity to communicate with muscle cells, what is now understood as the basis of nerve impulses. His original goal was to use electricity to cure paralysis. Galvani showed that by using two different metals connected in a series and attached to frog's legs, he could get the legs to move by jolting them with a spark from an electrostatic machine. In his experiments he used thousands of frogs (the "martyrs of science") to conclude that there was some fluid (now know to be electrolytes) that allowed bodies to restore movement. Galvani used electrical stimulation of frog muscle tissue to seemingly bring dead tissues back to life. This was eventually called, galvanism or "the science of electrifying dead bodies." By applying electricity to frog muscles he was able to show specific movement suggesting some artificial control of life's processes. In this respect Galvani can be con-

Frankenstein

sidered a real-life Dr. Frankenstein in attempting to restore life to the dead. Later, Volta showed that Galvani's frogs' legs could serve both as a conductor and a detector of electricity.

As noted above, Benjamin Franklin discovered the association of lightning to electricity and Alessandro Volta invented the battery (by putting alternating layers of zinc and copper together). And Galvani took this to another level when he used electricity to stimulate movement in frog legs. As a natural progression Galvani also used human cadavers and body parts, such as limbs and severed heads in his experiments. Upon stimulation by electricity, muscles contracted and relaxed causing diverse facial expressions, including opening eyes, flaring nostrils and the raising of eyebrows. Arms lifted up and legs moved. All of this hinted at the possibility of restoring life to dead bodies as well as body parts. The supposition is that if one can make individual body parts move then by stitching together body parts one could make an entire body move. (The problem, of course, is not movement but, rather, *directed* and *purposeful* movement!)

In one famous experiment Galvani had iron hooks installed on the roof of his house and during an electrical storm was able to transfer the lightning through his wire system to frogs' legs thereby getting them to kick (and confirming Ben Franklin's

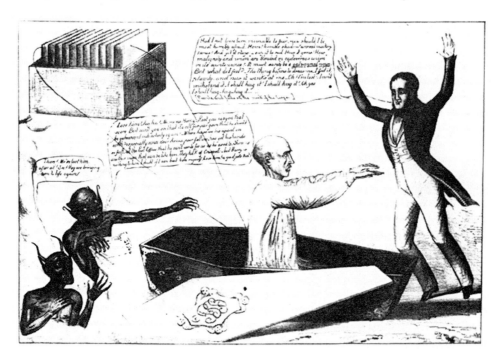

This image indicates that attempting to revive the dead was a popular concept of the early 19th century, contemporary with Mary Shelley. Note the presence of a "battery" (upper left corner) that is simply a series of metal plates (copper and zinc). Based on this image, it appears that the devil is in the details.

6. The Spark of Life

observation that lightning is electricity). As Galvani himself said of the experiment, "contractions not small occurred in all muscles of the limbs." Galvani used electricity via lightning to animate dead tissues, something Mary Shelley must have noted.

Using galvanism to animate dead matter was popular science during the late 18th–early 19th century, the time of Mary Shelley's growth from adolescence to early adulthood, and many a public demonstration showcased the effectiveness of how this worked. As an adolescent Mary herself was taken to such a demonstration by her father so she was certainly aware of the practice. As Mrs. Shelley herself said later, "Perhaps a corpse would be reanimated; galvanism had given token of such things."

A student of Galvani (and a nephew), Giovanni Aldini (1762–1834) went to London searching for a "perfect body," to experiment upon and test the limits of galvanism. Simply put, to reanimate the dead. At the time it was thought that if one could restart the heart then one could revive the dead. Though Aldini was able to show limited movement, his attempts at resurrection failed. Aldini can also be considered a real-life Dr. Frankenstein, even more so than his famous uncle, in trying to revive the dead.

In the 1818 first edition of *Frankenstein* Victor Frankenstein references galvanism when he realizes he must, "extinguish the spark which I so negligently bestowed." This is referring to galvanism in that by removing the spark the body parts would no longer move. This is simple logic in that since the body was animated with a spark then the removal of the spark would stop the animation.

In the Universal Studios' *Frankenstein* consider the bolts on the neck of the monster. One bolt is an anode (negative terminal, positive electrode) and the other is a cathode (positive terminal, negative electrode). In electrical current electricity enters via the anode and exits via the cathode and in doing so electricity enters the monster and animates and invigorates him. This would be the "spark of life." To reverse the process and destroy the monster, something Dr. Mannering attempts in the film, *Frankenstein Meets the Wolf Man*, would mean the drawing *off* of electricity thereby removing or reducing the electrical stimulus and ultimately extinguishing the spark. Think of a rechargeable battery in which the charge can be increased or reduced depending upon the flow of electricity. In *The Ghost of Frankenstein* an exponential increase in flow is invigorating when lightning strikes the monster. (Note: the lightning bolt only hits the neck bolt on the right, so this one is the anode and the one on the left is therefore the cathode.)

Mannering reads from Frankenstein's notebook, "Matter ages because it loses energy ... this my creation will never perish unless its energies are drained off artificially by changing the poles from plus to minus ... energy that can not be destroyed can be transmitted ... adjusting the plus poles to the minus poles will change the energy output of the nervous system." Ultimately, Mannering the scientist could not bring himself to drain energy from the monster and succumbed to the temptation to restore him to full power.

Anatomy and Physiology

Studies in anatomy date back to primitive man with cave paintings made during the Ice and Stone Ages. Anatomical studies became more serious with the ancient Egyptians who recorded them on papyrus. During the Reformation the Italians emerged as the anatomical experts creating texts with copious illustrations detailing the workings of the human body.

Andreas Vesalius's (1514–1564) book, *De Humani Corpus Fabricia* ("On the Fabric of the Human Body"), pub-

Image from *The Ghost of Frankenstein* where the Monster (Lon Chaney) is hit by a bolt of lightning that gives him renewed strength and vigor. Afterwards, Ygor exclaims, "Your father was Frankenstein and your mother was the lightning!" perhaps acknowledging Galvani, who used lightning to bring movement to animal limbs. The bolt of lightning only hits the right neck bolt so that one must be the anode and therefore the left neck bolt is the cathode.

lished in 1543, set the standard on anatomy for the next 250 years. The excellent anatomy illustrations helped inspire many famous artists from Michelangelo to Rembrandt in their anatomical renderings. In addition, the superb drawings of anatomy by Leonardo da Vinci helped to educate many generations with his accurate depictions of visceral detail. Such artistic anatomical illustrations influenced scientific advancements, as well, demonstrating that art does influence science. Another example is Bartolommeo Eustachius (1514–1574), an Italian anatomist who published, *Opuscula Anatomica*, in 1564, which focused on kidneys and the vascular system. Eustachius is considered one of the founders of the science of human anatomy (Eustachian tubes in the ear canal are named in his honor).

The invention of the printing press (in 1450 by Johannes Guttenberg) made anatomy books accessible to the public, which significantly increased general interest of the human body. During the 17th and 18th centuries anatomy flourished and essentially was the most prominent field of biology.

In Mary Shelley's time, during the early 19th century, anatomists and surgeons knew that the heart and brain were important, but their purpose and function were not fully understood. Anatomist Humphry Davy (1778–1829) showed that the heart pumps blood and is responsible for circulation, quite a remarkable discovery, as it was thought that the heart was where the soul resided. Davy was a supporter of the "vital powers" theory that gave man the ability to conquer nature. Attempts at reanimation through galvanism were the natural extension of this philosophy, which was discussed at the Godwin home where he was a frequent visitor.

Mary Shelley learned that human bodies were needed and used for the many anatomy demonstrations. In her novel, Victor Frankenstein says, "I collected bones from charnel-houses ... the dissecting room and the slaughter-house furnished many of my materials." Those needing cadavers were not particular whom or where they came from, as long as they were fresh (decayed bodies were of little use to the anatomist).

Many of the bodies used in demonstrations were the corpses of criminals. The real life use of criminal body parts being stitched together for galvanism experiments possibly inspired the idea of using a "criminal" brain in the 1931 *Frankenstein* film. Dissection meant that there would be no corporeal body for any sort of afterlife, so criminals feared being dissected and did their best to argue themselves out of that fate (few succeeded).

Alchemy

Alchemy originated in Egypt. The ancient word for Egypt, Khem, refers to the rich black soil of the Nile area. When Alexander the Great and the Greeks conquered Egypt they blended their culture of earth, air, fire, and water into the sacred science of Egypt (embalming and the afterlife) and the new word, Khemia, was used to describe this new fertile land with the blending of all elements of life. From Khemia we get chemistry. When Arabs conquered Egypt in the 7th century AD they called the "black land," Al-Khemia, and from that we get the word, alchemy. During the Crusades alchemical lore was introduced to Spain by the Arabian Moors and from there it gradually spread throughout Europe. The defining aims of early alchemy were the creation of the philosopher's stone, the ability to transform metals into gold, and the development of an elixir of life that brings eternal youth. Out of this mix of mythology, magic, and religion evolved the disciplines of chemistry and medicine.

In the 14th and 15th centuries some of the Arab alchemy writings were translated into Latin, which exposed many Europeans to this discipline making many converts and practitioners. With the influence of European interpretation, alchemists moved away from physics and devoted their study to mankind as being the alchemical vessel. This is an important concept that Mary Shelley used as a backdrop in her story, demonstrating how her knowledge of contemporary general science, including alchemy, helped to drive the basic plot of *Frankenstein*.

In the book, *Frankenstein*, Shelley actually mentions several alchemists by name, two of the most famous being Heinrich Cornelius Agrippa (1486–1535) and Theophrastus Bombastus von Hohenheim (1493–1541), better known as Paracelsus. Both of these men were key contributors to alchemical lore in Europe and helped to make alchemy the pop science of the day, through their influential books, along with others written by men such as John Dee (1527–1608). Paracelsus pioneered the use of chemicals and minerals in alchemical medicine and prompted the development of herbal medicines and plant remedies. Furthermore, his experiments with homunculus and attempts to make various forms of life were at the dawn of the medical and pharmaceutical branches. As a result of these alchemical efforts the workings of the human body were gradually discovered.

Body Snatching and the Resurrectionists

With the revolution occurring in anatomy, more cadavers were needed to satisfy the demand for teaching hospitals and medical schools as dissection material. At the time specific "demonstrations" were given on a particular aspect of anatomy and surgery and a fresh corpse was needed each time. With the rise in medical students and the subsequent need for cadavers, the practice of body snatching became prevalent. Body snatching was the sneaking into graveyards and digging up corpses, the fresher the better, and selling them to teaching hospitals for use in anatomy demonstrations. This method of obtaining bodies was controversial and there were many discussions during the early 19th century about the ethics of using bodies for anatomy demonstrations.

Those who dug up the bodies and sold them to medical schools for a living were called Resurrectionists. Those in the medical profession who obtained the bodies for demonstrations asked few questions. Later, those such as the infamous Burke and Hare, who dug up bodies for the medical profession, began to obtain bodies through murder to help satisfy the demand. Just so you gentle readers know, the exploits of Burke and Hare did not begin until the late 1820s, well after Mrs. Shelley penned her novel.

Volcanic Summer

The opening sequence in the 1935 film, *Bride of Frankenstein*, is based on reality. A young Mary Shelly (played by Elsa Lanchester) is induced to tell her tale of Henry Frankenstein, with a storm raging outside. The summers on the shores of Lake Geneva were typically calm and mild but not so in the summer of 1816.

The inclement weather began with the eruption of Mount Tambora, located on the island of Soembawa in Indonesia, on April 15, 1815. This particular eruption lasted a week and continued rumbling for 3 months. The Tambora volcano was originally 14,000 feet high and after the eruption the elevation dropped to 9,000 feet

6. The Spark of Life

(almost a mile decrease in elevation!). As a result about 10,000 people on the island were killed and another 80,000 would eventually die from starvation and diseases related to the eruption. It has been said that Tambora was one of the largest recorded eruptions with estimates of 1.7 million tons of dust put into the air, the equivalent of 6 million atomic bombs!

Due to the Earth's rotation and high atmospheric weather patterns, this volcanic dust reached the Northern Hemisphere during 1816, significantly blocking the sun thereby bringing about what has become known as the "year without a summer." For most of Europe the summer of 1816 consisted of inclement weather, many storms, and unusually overall harsh conditions.

Summary

The summer of 1816, the year without a summer, is when Mrs. Shelley wrote her book. So we can be thankful for a volcanic eruption that led to bleak summer conditions on the shores of Lake Geneva that helped put Mary in the mood to dream up the sage of Victor Frankenstein and his creation. The exploits of Mary and Percy Bysshe Shelley, Lord Byron, and Dr. Polidori at Villa Diodati in Switzerland during the dark summer of 1816 have been extensively written about and the unusual summer storms were an excellent source of inspiration for all of their "ghost stories."

All the work with galvanic electricity on animal tissues and the quest to reanimate the dead equally inspired the writings of Mary Shelley. So, during the late 18th and into the early 19th centuries the popular science of the time was galvanism, alchemy, and the new findings of the anatomists. All of this was on Mary Shelley's mind when, as a teenager, she created *Frankenstein*. The popular science of the time mixed with the results of the inclement weather changes in central Europe from the volcano and the suggestion of writing a ghost story all blended together in Mary's mind and thusly a legend was born. Each of these components was necessary for the final blend and with any of them missing then the result, the book, would most likely not have happened.

Mrs. Shelley took note of the popular science of her day and like a stew, mixed it all together, and created something that, 200 years later, still resonates and dominates many of our thoughts and "parlor" discussions. That spark has never been extinguished. An example of how long the shadow cast by Mrs. Shelley is: take note of the Universal Studios' Frankenstein films that are peppered with ideas and concepts touched upon by the popular science in her book, such as birth by electricity, invigorating by lightning, animating body parts, homunculi, and bringing dead tissue to life.

Now, 200 years later in the early 21st century, artificial life has been created at the cellular level. Stem cells are being driven to morph into full-fledged operational organs. A completely computer generated DNA life form has been created that can

Frankenstein

self-replicate, which would have been inconceivable during Mary Shelley's time. And yet, as awe-inspiring as these scientific advancements are, it is hard to imagine their fundamentals being stitched together in a manner that so evocatively brings to life the complexities of merging science and biology as did Mary Shelley's monstrous masterpiece.

7

Boris Karloff
The Walking FrankenDead

Due in large part to all of the 1950s and 1960s monster magazines I read growing up, I am a big fan of Boris Karloff. In watching his movies with the careful eye of the scientist, I have observed striking similarities between Universal Studio's *Frankenstein* (1931), and Warner Bros.'s *The Walking Dead* (1936). The similarities of these two films go well beyond standard plotlines or the inimitable presence of Karloff in a starring role.

Decades before Robert Kirkman's wildly popular comic book *The Walking Dead*, and the very popular TV show on the AMC channel, there was the 1936 Karloff film, *The Walking Dead*. The premise of someone coming back from the dead to take care of old business, namely revenge, is a common one. In *Frankenstein* the monster wants revenge on his creator and in *The Walking Dead* John Elman seeks vengeance on the racketeers who framed him for murder. Beyond the desire for revenge another, far more interesting plot point, ties these two movies together. In both films Karloff's characters, the monster in *Frankenstein* and Elman in *The Walking Dead*, are resurrected via electricity. It seems that with Karloff, electricity and revenge go hand in hand.

1930s Biomedical Science

As mentioned, *Frankenstein* was filmed in 1931 whereas *The Walking Dead* was filmed in 1936. Though *The Walking Dead* had slightly more interesting science the biomedical differences in the elapsed 5 years from 1931 to 1936 were not that significant in major advances with one notable exception (see below discussion on Lindbergh heart). Though the lab setups are contemporary in their setting in both of the films, it should be noted, however, that in *Frankenstein* the revival by electricity closely resembles the knowledge and understanding of Galvanism of the day whereas in *The Walking Dead* the use of electricity has a slightly different approach and meaning. Furthermore, as discussed in detail below, the use, in 1936, of the just developed "Lindbergh heart" in the film, *The Walking Dead*, is an excellent example of using then contemporary 1930s science for plot devices. Also, heart-lung machines to supply oxygen to the brain during surgery were being developed during this time so this too is contemporary with *The Walking Dead* film.

Frankenstein

What is Death?

After the demise of the Wicked Witch of the West in the popular film, *The Wizard of Oz*, the coroner proudly croons, "And she's not only merely dead, she's really most sincerely dead." How dead is dead and when are you "really most sincerely dead"? Even in the 1931 film, *Dracula*, the Lugosi character intones, "To be really dead. That must be glorious!" Or as Dr. McCoy of the original Star Trek TV series says, "He's dead, Jim!"

But what does that mean? There are drowning victims that have been declared dead, but were revived sometime later. Some of these victims were without oxygen for hours so the brain would be affected. However, when drowned in cold water the lower temperatures provided a protection to the delicate brain tissue. The human body can tolerate many extremes yet still able to be revived. Therefore, this then begs the question: how do we define dead?

Dying begins when the body does not get enough oxygen to survive. The earliest description of death, in the first edition of the *Encyclopedia Britannica* (1770 edition), was, "the separation of the soul and the body," which clearly associated death with religious beliefs. Death is not an easy thing to define and with modern medicine and technology the determination of when a person is "really most sincerely dead" is difficult. Even with the obvious signs of missing respiration and heartbeat there is still the chance of the person still being alive. (Think of Edgar Allan Poe and "Premature Burial," the protagonist buried and entombed before being dead.) And there is technology that can make some forms of death reversible. A ventilator can keep someone artificially breathing; feeding tubes can keep someone hydrated with proper nutrition; CPR (cardiopulmonary resuscitation) can be used to revive the newly dead. However, there is more to life than a pulse since there are some patients that do not revive even after being hooked up to medical machines. A coma is one such example, not yet dead though not totally alive either.

In 1968, physicians at Harvard Medical School defined death as an irreversible damage to the brain or brain death. This implies a person's memories and personality, located in the brain, define life, and when they are gone that person is dead, irrespective of whether other body functions continue. This also raises the question of how much of the brain has to be dead to be declared dead, a problem many bioethicists and doctors are considering.

How Death Occurs

There are many, many ways death can occur. Some are accidental, some violent, and others natural. And even among these categories there are many combinations, all of which affect the likelihood of whether the body can be revived. In one scene in *Frankenstein* the good doctor and his assistant cut down a body from a gallows and determine "it's no good" since the neck is broken, implying the brain is useless.

7. Boris Karloff

In an earlier scene the doctor and Fritz exhume a freshly buried coffin for the body with the understanding that the fresher the better. Though a head may be damaged, with respect to Dr. Frankenstein, then the other body parts may be useful. In the case of John Elman in *TWD*, his revival began as quickly as possible after his electrocution, perhaps just a few hours after the event. No matter how death occurred in these two films the body was brought back to life by electricity.

Just to keep you gentle readers up to date there are some other considerations about the nature and matter of death. Such examples are suspended animation and the so-called cryonauts (deep-freeze future). They want their newly dead bodies to be cryopreserved so when future technology is available they can be revived. Again, this forces us to rethink what we mean as dead. In many respects cryonauts want to achieve some sort of immortality. In addition to preserving whole bodies attempts have also been made to preserve organs, tissues, and heads. The hope is that sometime in the future death will be conquered.

Different cells in our bodies die at different speeds so the actual dying process can be quite lengthy. The organ that requires the most amount of oxygen to survive is the brain and when deprived of oxygen from three to seven minutes brain damage can occur. When the heartbeat and breath stop the person is clinically dead. However, it should be noted that clinical death is reversible and that clinical death is not necessarily brain death. The point of no return is biological death which begins around four to six minutes after clinical death.

What Happens to a Body After Death

All bodies undergo changes after death, some more than others depending upon a wide variety of circumstances. And there is a specific sequence by which change occurs. Change begins at the molecular level and then progresses to microscopic and eventually to gross morphology. Two main processes, putrefaction and autolysis, immediately begin to alter the body; either one may predominate, depending on the circumstances surrounding death, as well as the climate. Putrefaction involves the action of bacteria on the tissues of the body. This process, prevalent in moist climates, is associated with green discoloration of the body; gas production with associated bloating; skin slippage and a foul odor. Autolysis is the breakdown of the body by endogenous substances and proceeds most rapidly in organs such as the pancreas and stomach. In most instances both putrefaction and autolysis occur simultaneously and in temperate climatic conditions they can result in rapid degradation of the tissues and body. With the pancreas destroyed then no more insulin will be made effectively making the body diabetic. Upon revival, if the pancreas is still damaged, then the body will suffer from diabetes, which may explain why some revived creatures are so angry and in pain; their hormone levels are out of balance.

Within an hour of death all of the body's muscles initially relax (called primary flaccidity). The eyelids relax, the pupils dilate, the jaw may relax and open, and all body joints are flexible. When all muscles loosen then the skin will sag making some bone structures more pronounced. When the heart stops beating the circulatory system stops and the blood pools and settles by draining from small veins and arteries (the average human has about 5 to 6 liters of blood, about 6 quarts). Along with this the body will begin to cool from its normal 37°C degrees (98.6°F) until reaching room temperature. The cooling of a human body is down two degrees Celsius in the first hour and one degree each hour afterwards. The rate of cooling varies depending upon body location, such as shade versus sun, clothing worn, and temperature. A body would cool more quickly after dying outdoors during the winter compared to one dying in a desert. Fat or obese people lose heat slower than infants, who cool rather quickly mostly due to their smaller muscle mass. Otherwise, it takes the body about 24 hours to completely cool, or become the same temperature as its environment.

From the second to the sixth hour after death more settling of the blood occurs where parts of the body nearest the ground have a deeper purple color compared to surface skin. During this time chemical changes occur within the body and the muscles begin to stiffen, a process called rigor mortis (caused by a complex chemical reaction involving lactic acid and myosin), which forms a gel like substance that stiffens the body. Not all muscles stiffen simultaneously. The muscles in the eyelids, jaw, and neck stiffen first followed by those in the face and down the chest, abdomen, arms, and legs. The last to stiffen are the fingers and toes.

During the hours 7 to 12 after death there is the maximum rigor mortis stiffening of muscles throughout the body. The knees and elbows may be slightly flexed and fingers and toes can appear oddly shaped and pointed. Beyond 12 hours after death the chemical changes continue and the muscles begin to relax and loosen. Internal decay begins from the first to the third day, though this is highly dependent upon external conditions such as temperature. Due to blood settling, parts of the body will appear blue within 8–12 hours after death.

Though the body as a whole may be dead, some things within the body are still alive. Some skin cells can be harvested up to 24 hours after death and all the gut microbes will be alive for some time. It is these intestinal microbes, along with their enzymes, that cause decomposition of the body. The body becomes discolored, first turning a green, then purple, then black. The odors given off by decomposing flesh are quite scary indeed! Decomposition reactions gives rise to gas which can cause a body to bloat, the eyes bulge out of their sockets, and the tongue to swell and protrude.

As noted, decomposing starts almost immediately after death with the skin going through several changes in color. Since different cells die at different rates (e.g., brain cells die within a few minutes while skin cells can survive over 24 hours after death) the rate of decomposition can initially be somewhat slow. A week after death the skin blisters and can easily fall off. Within a month the hair, nails, and teeth can fall out.

7. Boris Karloff

Internal organs and tissues will have liquefied and eventually evaporate, leaving only the skeleton.

Some think that fingernails and hair continue to grow after death, but this is a myth. What actually happens is that the skin dries outs and pulls away from the nails and hair which makes them more prominent, giving the illusion of growth. Some of the last cells and tissues to undergo decomposition after death are tendons and ligaments.

Decomposition in the air is twice as fast as when the body is under water and four times as fast as underground. A corpse left above ground is rapidly broken down by insects and animals, including bluebottles and carrion fly maggots, beetles, ants and wasps. A corpse can become a moving mass of maggots within days, even hours in hot climates. Approximately 150,000 maggots can be found on an exposed corpse.

It can takes decades for a body to decay, as there are many factors that affect the rate of decomposition, such as whether the person was embalmed or not, what type of casket and vault they were placed in (if any), humidity, heat, cold, soil type, water level, depth of burial, the availability of oxygen, accessibility by insects or scavengers, body size and weight, clothing, the surface on which a body rests—all determine how quickly a fresh corpse will skeletonize or mummify. A basic guide for the effect of environment on decomposition is given as "Casper's Law" which determines that where there is free access of air a body decomposes twice as fast than if immersed in water and eight times faster than if buried in earth. People who have been dead for decades could still look fine while others of the same era are completely decomposed.

Brain Death

The brain controls all body functions and yet there are three things the brain cannot do. The brain cannot feel pain, the brain cannot store oxygen, so it must constantly be replenished, and the brain cannot store glucose, so it too must constantly be supplied. The brain can survive up to about six minutes after the heart stops.

Those who are brain-dead show no response to command; the patient has flaccid or nonresponsive arms and legs; the pupils are unresponsive or fixed (no response from the optic nerve); no corneal reflexes meaning no blinking when irritated; no gag reflex; and no spontaneous respiration. Even though the patient may be brain-dead there may still be some spinal cord reflexes such as knee jerks when pressure is applied to the hand or foot. Patients who suffer from brain death are not in a coma and those patients who are in a coma may or may not become brain-dead. Those in a deep coma are considered to be in a vegetative state.

A brain can be injured by either natural causes or by trauma and three possible results can occur: bleeding, swelling, or both. Brain death by trauma can be either open (gun shot) or closed (blunt injury). There is also brain death by anoxia (no oxygen such as by drowning, smoke inhalation), by a cerebral vascular accident (stroke, aneurysm, infection), by a tumor, or by a drug overdose.

Frankenstein

Necrosis

Necrosis is caused by factors external to the cell or tissue, such as infection, toxins (such as venoms), or trauma (either physical such as with extreme temperatures or damage to blood vessels) that result in the unregulated digestion of cell components. In some cases the digestion and release of cellular components can result in inflammation that can prolong the necrosis as seen in extreme cases such as gangrene. Necrosis begins right after cell death.

There are five types of necrosis. Coagulative necrosis is the formation of a gel-like substance in dead tissues primarily caused by protein degradation and mostly observed in kidneys, heart, and adrenal glands. Liquefactive necrosis is a viscous mass, commonly called, "pus," composed of digested dead cells and typically comes from fungal or bacterial infections. Caseous necrosis is a combination of coagulative and liquefactive necroses and the dead tissues look like clumped cheese. Fat necrosis is when lipases in fatty tissues degrade dead cells. Lastly, fibrinoid necrosis is caused by immune response related vascular damage by complexes of antibodies and antigens that get deposited on the walls of arteries.

Resuscitation

Resuscitation is bringing someone who is unconscious, not breathing, or close to death back to a conscious or active state again. Resuscitation primarily involves restarting the heart using a combination of cardiopulmonary resuscitation (CPR) and an electrical shock, or a defibrillation. Also, it should be noted that CPR alone is not enough to restart the heart. The goal of CPR is to help preserve brain function until further measures are taken to restore spontaneous blood circulation and breathing in a person who is in cardiac arrest. To get enough oxygen the heart must pump blood into the brain, if not naturally then mechanically. Blood circulation and oxygenation are required to transport oxygen to the brain and other tissues. The main objective of resuscitation is to delay tissue death.

Reviving the Newly Dead (or the Long Dead)

To revive the dead there are a few things that are important. First of all, the heart needs to beat so blood can be circulated. Without circulating blood then no oxygen will get to the brain and therefore, no reviving. The brain must be sufficiently cohesive that some normal physiology and function are present. The brain makes most of the important hormones in our bodies and this function must remain intact. In addition to oxygen there are also general nutritional needs that must be addressed. You need fuel to make the body go. Though not particularly necessary it would be most helpful if all of the senses were properly working. You can certainly survive without taste, smell, or hearing, but it would be difficult. And the sense of touch would be useful too. A blind monster would not be especially useful (for example, the end scenes of Universal's, *Ghost of Frankenstein*, where the monster went blind due to an autoimmune reaction from the transplanted brain). For general movement the muscles should be in good working order as well as the immune and digestive systems. Overall, normal physiology of a normal working body would be necessary for revival.

Frankenstein *(1931)*

After Henry and his assistant, Fritz, secure the stitched together creature to a gurney and raise him up into the sky for his electrical reanimation, potentially thousands of volts of lightning pass into the monster's body. The electrical gadgets visible in the lab, consisting of various wires, capacitors, rheostats, resistors, and amplifiers process the electrical bolts into the monster's body. However, it is unclear how much electrical juice, both voltage and amperage, actually enter the monster. Once the heart had started, life had been bestowed, so no further electrical intervention was needed. The heart is key, since it would bring much needed oxygen to the monster's brain, albeit a criminal one.

Regarding the work of Dr. Frankenstein as a student, his mentor, Professor Waldman said, "his researches in the field of galvanism and electrobiology were far in advance of our theories here at the university. In fact, they reached a most advanced stage. They were becoming dangerous."

As Dr. Frankenstein is about to conduct his ultimate experiment he credits, "all the electrical secrets of heaven," before the moment of reanimation, acknowledging the importance of electricity. Once the bolt of lightning hits the instruments, huge sparks and electrical flashes indicate that all of the assembled apparatus are being overridden by the strength of the lightning bolt. And the lightning serves to jolt the stitched together body to life, prompting Dr. Frankenstein to maniacally shout, "It's alive!, It's alive!," and then to blasphemously say, "In the name of God, now I know what it is like to be God!" Bringing the dead to life.

What is unknown is how long each of the monster's body parts have been "dead." No doubt some parts were dead before others. For example, Dr. Henry secured the main torso and, based on the visible scars, stitched on new hands. Which died first, the body or the hands? Most likely the main torso died first and all other body parts, be they limbs or internal organs, were added later. Apparently, the last item to be added was the brain. If the various body parts were dead for some time then internal decomposition had started with some necrosis visible. Perhaps Dr. Henry had replaced all internal organs with newly dead ones to eliminate decomposition issues. Hopefully, he had some sort of ice box nearby so he could better preserve the delicate tissues as he assembled his creation.

Frankenstein

The Walking Dead *(1936)*

Warner Bros. took advantage of the star power of Boris Karloff, hot off the recently released *Bride of Frankenstein* (1935). John Elman (Karloff) is a musician who is sent to the electric chair after being wrongly accused of murder. His newly dead body sparks the interest of Dr. Evan Beaumont (Edmund Gwenn) a cardiologist working for "Medical Sciences Foundation Research Laboratories."

Immediately after learning that Elman was killed by electrocution, Beaumont asks the Governor to call off the autopsy, knowing that he is "without a second to lose." It should be noted that Beaumont understood the time constraints of working with the newly dead since each minute that lapses is one more minute of decay and putrefaction of a body so he wanted the body immediately after execution before too much irreversible decay occurs. Elman was electrocuted at midnight and within a few minutes the autopsy was cancelled. At this early stage after death, only minimal rigor mortis has set in and there is little, if any, putrefaction and autolysis. The time from his electrocution (midnight) until he is placed on a heart-lung machine (early morning) was no more than a couple of hours, so there is no decay. However, some blood pooling should have occurred since the body was laying flat right after the electrocution as it was being transported to Beaumont's lab. Beaumont needs the untouched body to conduct his resuscitation experiments.

Regarding Beaumont's revival plans an attending nurse says, "Its impossible (to revive Elman). He was electrocuted." Meaning his brain was fried. Once in Beaumont's lab early that morning, Elman was immediately put on a heart-lung machine to keep oxygen pumping to his brain. This would have taken care of the blood pooling issues and could have delayed if not stopped the putrefaction steps. Reversing what putrefaction did occur would require active metabolism, meaning a resuscitated Elman. Considering the time involved Elman was most likely without oxygen for about an hour, depending upon how quickly the body was transported to Beaumont's lab, before he was placed on the heart-lung machine.

Electrical Stimulation

Resuscitation of a man killed by electrocution would certainly be difficult, primarily due to the severe destruction of not only the brain, but also the heart. Dr. Beaumont attempts to revive Elman via surges of electricity pulsed through a metal skull cap (similar to that from the electric chair), metal clamps on each wrist, and ankles, and metal tubes covering each finger. The finger tubes serve the same purpose as Frankenstein's neck bolts; they are conductors of electricity.

While on the gurney an acrylic box helmet-like covering is placed over Elman's head. Most likely this served as a pseudo hyperbaric chamber that delivered a high concentration of oxygen to help the revival process. A respirator cup placed over

7. Boris Karloff

the mouth would be just as useful in delivering oxygen though nowhere near as dramatic.

Elman's gurney slowly rocks back and forth, like a teeter totter, during the electrical stimulation. The gentle rocking could be the doctor's attempt to help blood circulation and bring oxygen to the brain, as well prompt the heart to begin beating. Furthermore, the gentle rocking could help move body fluids around to assist the heart to begin steady beating. At the least the rocking could help dissolve the blood pooling seen with the newly dead. In some ways, this rocking could be seen as an early example of CPR, the primary purpose of which is to keep oxygen flowing to the brain. No such rocking was seen in *Frankenstein,* so perhaps the power of a lightning bolt was sufficient to shock his heart into beating.

"It's Alive!"

Sure enough, the sustained rocking causes Elman's heart began to beat again. After the successful operation, Dr. Beaumont makes the declarative statement, "It's alive! He will live." [Note the use of the word, "It."] Then a radio announcer states, "Modern miracle performed by Dr. Beaumont" followed with the comment by a reporter, "The most incredible achievement in the history of medical science." The banner of a newspaper states, "Elman 'brought back' to life!" with the headline: "Science baffled as electrocuted man lives again." Dr. Beaumont refers to Elman as "The man who returned from the dead." In terms of timing, one newspaper headline says it all, "Electrocuted last night, lives again today."

It should be noted, that in some scenes, particularly those filmed after Elman has been brought back from the dead, Karloff is wearing lifts in his shoes thereby making him look taller. And, according to the Warner Bros. Pressbook, a man's body grows about an inch longer after electrocution. Of course, pressbook "facts" should always be looked at with a skeptical eye if not two skeptical eyes. Or, if you are like "Marty the mutant" from the film, *The Day the World Ended,* then you can look with *three* skeptical eyes!

After being revived Elman is lethargic, indicative of a not quite fully functional brain and perhaps a few stiff muscles. He exhibits symptoms of amnesia. Unable to remember his name or why he is there. Perhaps his brain was deprived of oxygen for too long from the time of electrocution to revival. His overall demeanor is one of slow passiveness. This is not especially surprising since an electrocuted body would not have many unaffected organs. In reality, beyond a compromised brain, most of the internal organs, including the heart, would be completely fried.

But back in the realm of *The Walking Dead,* Beaumont asks the revived Elman, "You were in another world … do you recall anything of that world before you came back?" Beaumont seeks to understand the impact of death on the brain. Elman is unable to answer Beaumont's questions. However, in spite of his limited mental capac-

Frankenstein

ity, Elman has enough sense to seek out those who framed him to make sure he got his revenge.

The Laboratory of The Walking Dead

Dr. Beaumont works at the Medical Sciences Foundation Research Laboratories. His colleague, Jimmy, while looking at a cardiology experiment, comments on Beaumont's heart research, "Imagine Dr. Beaumont keeping this heart pumping for over two weeks." The heart appears to be a cow heart. Quite a bit of cardiology research was performed on cow hearts during the early 1930s, so the writers and producers were contemporary in this detail of the film.

Beaumont's laboratory at the Foundation is a nice setup with two large, well-equipped rooms full of copious glassware, multiple racks of test tubes, bottles of chemicals, microscopes (with one under a glass bell jar), and other assorted bench bling. Some of the glass vessels were plugged with cotton, indicating they were sterile and most likely contained biological samples. Along with the aforementioned heart-lung machine are elaborate electrical apparatus and machinery. In the lab is an unusual device that shoots off large amounts of steam; its purpose unknown. Visible on the lab benches are nice microscopes, an autoclave, gas cylinders, several shelves of chemicals, many of which are in dark glass containers to reduce exposure to light.

Also present is a glassware wash station, something rarely seen in any SF lab. Later, we see a small heart in an elongated glass vessel that appears to be beating. The many chemicals and elaborate glassware are suggestive of working on small molecules and simple-to-purify reagents. There are also several glass tanks and vessels filled with liquids, most likely buffers and reagents to supply nutrients for the heart.

From the film *The Walking Dead*, the laboratory of Dr. Beaumont where lab assistants Nancy (Marguerite Churchill) and Jimmy (Warren Hull) are monitoring a beating cow heart.

7. Boris Karloff

Charles A. Lindbergh and the Artificial Heart

During the resuscitation procedure an elaborate lab setup is seen where a primitive heart-lung machine is used to stimulate the heart and provide much needed oxygen to the brain. Multiple pumps and liquid transfer vessels interconnected with rubber tubing are in the background. As colleague Jimmy says, "Keep that Lindbergh heart pulsating. See that it doesn't stop."

Charles A. Lindbergh (1902–1974), known for his transatlantic flight in May 1927 was also interested in cardiology. His sister-in-law was fighting mitral stenosis (narrowing of the mitral heart valve). Lindbergh, along with Nobel Prize winning Dr. Alexis Carrel at the Rockefeller Institute, actually helped to invent an artificial heart. Lindbergh and Carrel developed a system to keep organs alive outside of the body by circulating nutrient-rich fluids through them. They eventually perfected a glass perfusion pump that could maintain a heart in a sterile environment. This breakthrough helped other scientists eventually create the first artificial heart. The film, *The Walking Dead*, was produced in 1936, so the writers referring to the "Lindbergh heart," were very up to date with biomedical advances and quickly incorporated such findings into their scripts.

Two ways are seen showing the newly beating heart of Elman. The first is via glowing neon lights in the shape of a heart and the second viewing is through a chest x-ray. Chest x-rays are static images like a photograph and showing a beating heart on a static image is impossible. Also, the exposure time of an x-ray is short, too short to record the beats of a heart. In reality, a fluoroscope would be used to see a beating heart in real time. The use of a mechanical heart by Dr. Beaumont to keep pumping oxygen to the brain does foreshadow a time when such procedures would be common. Mechanical hearts are now routinely used to keep a patient alive while their heart is being operated on. These mechanical hearts pump blood through the body thereby maintaining essential physiology and delivering oxygen and nutrients to the brain.

From the film *The Walking Dead*: a small heart (from a rabbit?) beating in a Lindbergh chamber. In the film the fluid is seen flowing over the organ, providing nutrients and oxygen to sustain the tissues. The forced flow of vital nutrients is called perfusion.

Frankenstein

Blood Clot Amnesia

An x-ray shows a blood clot in the hindbrain of the revived Elman and while examining the image, Beaumont says, "Except for this blood clot, Elman is apparently sound." He is unsure as to why Elman has amnesia and is so lethargic in his movements. A blood clot would be very difficult to detect on an x-ray and not as obvious as it appears in the film. It should also be noted that an x-ray of the skull would not be particularly revealing about brain function. There are other tests such as an EEG (electroencephalogram) that would be much better in relaying the condition of Elman's higher brain functions. The implication is the bloodclot is the cause of the amnesia.

At a piano recital for Elman, Beaumont introduces his newly revived "walking dead" patient as "the man who returned from the dead." After dispatching a few of the scoundrels who framed him for murder (as Beaumont says to the district attorney, "taking justice in his own hands") Elman ends up at the Jackson Memorial Cemetery. When asked why he is there he responds, "It's quiet. I belong here." At the cemetery the two remaining racketeers who framed Elman shoot him 7 times with a 6-shooter. While dying Elman solemnly instructs Beaumont, "Leave the dead to their maker. The Lord God is a jealous God."

One is tempted to interpret Elman's seemingly haunting post-revival performance after being resuscitated as somewhat unhappy and hollow since he has seen Heaven when dead and was taken away from there without his permission and brought back to Earth. Elman longs to return and dying again would be the quickest way there.

The Frankenbrow and Other Physical Similarities

Several scenes in *The Walking Dead* show John Elman looking very much like the Frankenstein monster. The various camera angles, close-ups, and the way Karloff's face is lit all recall scenes from *Frankenstein*. In a profile shot it is difficult to tell the difference between the monster and Elman. You can not mistake that brow! One wonders if this is Karloff's natural forehead or was it perhaps enhanced to not so subtly resemble the Frankenbrow? Tellingly, his profile is notably more pronounced in *The Walking Dead* than in his other non-Frankenstein films.

Another physical anomaly of Karloff's that was skillfully exploited was a denture plate the actor wore on the right side of his mouth. This was taken out to gruesome effect in *Frankenstein* and *The Walking Dead*. After he is newly brought back from the dead and still on the gurney, the camera pans a close-up of Elman's caved in right cheek, and his heavy breathing causes the cheek to dramatically swell and collapse, resulting in a ghastly visage.

After being revived from the dead, Elman sports a streak of gray hair on the top

7. Boris Karloff

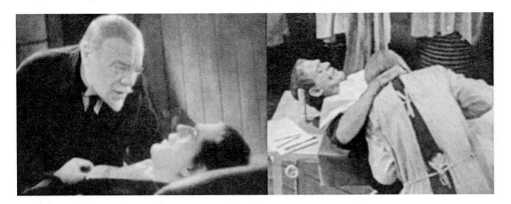

A composite image of Karloff from both *The Walking Dead* (left image, with Edmund Gwenn) and *Frankenstein* (right, with Edward Van Sloan) clearly showing that distinctive Frankenbrow.

right of his head, which is easy to interpret as a nod to *Bride of Frankenstein*. A comparison of the dialog also reveals a rather morbid connection between the two films. At the end of *Bride of Frankenstein*, Karloff's monster, with his hand on the lever, says "We belong dead" and in the film, *The Walking Dead*, Elman says, "I belong here," referring to a cemetery.

Warner Bros. vs. Universal

The Universal films, *Frankenstein* and *Bride of Frankenstein*, were filmed in 1931 and 1935, respectively and the Warner Bros. film, *The Walking Dead*, was filmed in 1936, so the difference in time of production is minimal. The Universal films depicted electricity from lightning, a natural source impossible to control, but in *The Walking Dead* the electricity was man made, and therefore more receptive to human influence.

It is interesting to note that the music for *Frankenstein* and *The Walking Dead* were arranged by the same composer, Bernard Kaun. Also, the music score for Elman's electrocution revival interestingly mimics the creation score for *Bride of Frankenstein*. The piano recital that Elman gives after he is revived is Anton Grigorevich Rubinstein's "Kamenniy-Ostrov."

Another interesting consideration is how the films approached the theme of death and revival. In *The Walking Dead* the objective was not to defeat death, but rather, to understand the "other side." Whereas in the Frankenstein films, the intention was indeed to defeat or even to cheat death. No matter the difference, Boris Karloff is indeed the Walking FrankenDead.

PHYSIOLOGY

The study of normal vital processes is called physiology. What binds all life forms together is their dependence upon a few simple physiological processes. All life forms, from simple bacteria to complex mammals, including man, must obtain nutrients and eliminate waste products. A key component to this is oxygen and the ability to use energy. And since environments are constantly changing, an organism's physiology must adapt as well. How these challenges are met, from maintaining internal homeostasis to responding to external stimuli, vary from species to species, and, of course, monster to monster. Body size, shape, and metabolic rate, often regulated by drugs and hormones, are dramatically affected by alterations in the environment. While the outside environment can change, the body's internal components must stay the same for survival.

8

The Hairy Who Are Scary

Hair. A blessing, a curse, it's in the way, too much of it, not enough of it, can't do anything with it. Hair seems innocuous enough and maybe even potentially boring, but it really is a fascinating subject, especially in SF films, where it often takes on a life of its own. We will be focusing on those movies with creatures burdened with hair that comes and goes, such as werewolves, and not with creatures permanently hairy like Yeti or Bigfoot.

Hairy Info

Humans have about 5 million hair follicles on our bodies and *all* of these are formed by 22 weeks in a developing embryo. There are about 1 million hair follicles on the head, and about 100,000 are on the scalp. Follicles are never added during life, so you are born with the most you will ever have. As body size increases during growth the density of the hair follicles in the skin decreases.

One of the defining characteristics of mammals is their hair. Animals, including humans, have many different types of hairs on their bodies. For humans, hair is on most areas of the body with the most obvious, of course, being the head and face. Areas completely devoid of hair are mucous membranes and areas called glabrous skin such as the tips of our fingers, palms of hands, lips, and soles of feet. For humans most hair is located on the ears, face, eyebrows, armpits, scalp, legs, and pubic region.

Of the 5,000 or so species of mammals just a handful do not have any fur. Those without fur, including man, are elephants, hippopotamuses, rhinoceroses, pigs, walruses, and cetaceans like whales, not to mention naked mole rats. Most mammals have a light skin color covered with fur. Only horses and humans are capable of sweating over most of their bodies.

During the evolution of man, from furry creature of *Australopithecus afarensis* to near naked current *Homo sapiens*, hair went from being straight, as seen with just about all mammals, including apes and chimpanzees, to being curly and kinky. This type of hair protected the scalp and therefore the brain, from harmful UV rays from the high equatorial (African) sun. As man migrated to colder climates the ability to grow long, straight, densely packed hair provided a distinct evolutionary advantage

8. The Hairy Who Are Scary

Cover to *Scary Monsters* #93 featuring the article "The Hairy Who Are Scary." ©2016 Dennis Druktenis Publishing & Mail Order, Inc.

and would be a distinct disadvantage in a hot climate when compared to loosely packed, spongy, closely cropped hair. As a result, those near the equator have curly hair to protect from the sun's rays whereas those in colder climates have denser hair to keep the head warm.

Structure of Hair

Hair is a filamentous material primarily composed of the protein keratin that grows from follicles in the skin. These follicles in our skin produce the two most common types of hair: terminal and vellus. Mature, thick hair is called terminal, while the fine hair growing over the body surface, as on the forehead and abdomen, is called vellus. Simply stated, terminal hair is thick whereas vellus hair is thin and fine. Either way, hair is composed of two components, the bulb (what you see at the end of the hair shaft when pulled from skin) and the shaft, which is the part that extends above the skin (and what you comb and cut). The bulbs are embedded in the skin and regrow hair, via stem cells in the bulb, after the hair follicle has fallen out. The bulb contains the cells that produce the hair shaft so hair growth actually begins inside the hair follicle. Also, due to the rapid growth of hair, the follicles have increased metabolic rates.

Physiology

The hair shaft itself is composed of three parts, the medulla, the cortex, and the cuticle. The external cuticle consists of several layers of thin, flat cells overlapping each other like roof shingles; the middle cortex contains the keratin bundles and helps provide rigidity to the hair shaft as well as water transport; the inner medulla is a disorganized area that is not always present in each hair fiber. The cortex contains melanin, the pigment which gives hair color. The hair cortex is produced by keratinization of cells in the hair bulb, and once keratinization has occurred, no amount of trimming of cutting or bleaching on the surface will have any effect on the rate or thickness of growth.

The shape of the follicle controls the shape of the cortex and its shape is determined by how straight or curly the hair is. Asian hair has a round fiber and is very straight whereas oval or irregularly shaped fibers are more wavy or curly as often seen with those of European and African descent. Each hair shaft is naturally covered with a single molecular layer of lipid, from the lubricating sebaceous glands near each bulb that help repel water. The diameter of each strand of human hair varies from 17 to 180 micrometers (0.00067 to 0.0071 inches). The only "living" portion of hair is the follicle. The hair strand after keratinization has no metabolic activity and as such is considered "dead."

The growth of hair is cyclic and proceeds through three distinct phases. The prolonged growth phase is called anagen and the short involution phase (hair follicle death) is called catagen when the hair follicle shortens and an anchored club hair is produced. The rest phase (hair shedding) is called telogen. If the anagen phase is prolonged excessive hair growth will result. All three phases can occur at the same time with one strand of hair being in the anagen phase with another being in the telogen phase. As mentioned, humans are born with approximately 100,000 hair follicles in the scalp. Each day, approximately 100 hairs are shed from the scalp and about the same number of follicles enter the growth phase. It is the duration of growth that determines the length and volume of hair. The growth and rest cycles are highly regulated by complex interactions between cells in the skin layers.

Hair on the body is called vellus hair, which is short, fine, and non-pigmented. These vellus hair follicles can become larger or smaller under systemic and other local influences that alter the length and time of growth (anagen) and the individual hair volume. Some androgens (androsterone and testosterone) do regulate vellus hair growth, primarily in the face (beard), chest, and limbs. Men have more body hair than women, and more testosterone, so men have more terminal hair, especially on the face, chest, back, and abdomen. Women have more vellus hair.

A third type of hair is lanugo, very fine, soft and unpigmented hair that is typically found on the bodies of newborn babies. Lanugo is the first hair to be produced by the fetal hair follicles and usually disappears shortly after birth. After lanugo hair is shed it is replaced by vellus hair.

The angle of growth of hair is coordinated in the same direction on various parts of the body. This is important because the hair angle on humans is different than on wolves so for a werewolf to exist his hair growth would probably have to change directions. Not an easy thing to do and may explain some of the pain seen with various film transformations. The rate of growth and type of hair are unique for specific body areas. Hair on the scalp is programmed to grow long, whereas that on the arm is rarely longer than one to two centimeters.

Function of Hair

For mammals hair serves many functions, including warmth and protection. It provides thermal regulation (heat insulation), and protection in the form of camouflage, as well as a means to communicate through signals and displays, either relaying a warning to a potential rival or an invitation to a potential mate. Hair also has a sensory function that extends the sense of touch beyond the skin's surface as well as protecting the head from UV radiation.

The hairs on human skin normally lie flat during hot conditions. The arrector pili muscle, those muscles that cause hairs to stand up, relaxes allowing the hair to lie flat and this prevents heat from being trapped by a layer of still air between the hairs. In other words, hair lying flat allows heat loss through convection. Also, when a human body is too cold the arrector pili muscles bound to hair follicles contract, called piloerection, making the hair strand stand up which then forms a heat-trapping effect immediately above the skin surface. The effects of piloerection are more commonly known as good old fashion goosebumps. Hair also provides a cooling effect as sweat evaporates from soaked strands. In other mammals, especially wolves, the fur fluffs up due to piloerection that helps to insulate the body from the cold. When a body is too warm the opposite occurs in that the arrector muscles make the hair follicle lay flat on the skin which helps heat to escape more readily.

Different types of hair on the body serve different functions. For example, eyebrow and eyelash hair provide mild protection to the eyes from the elements (dirt, rain, sweat). Eyebrows are also involved in non-verbal communications and can display various emotions like anger, excitement, sadness, and surprise. For most mammals, eyebrow hair is much longer with whisker-like hairs that are mostly tactile sensors. When hair shafts move, either through touch or air, by displacement or vibration, nerve receptors within the skin sense this with eyelash hairs being especially sensitive to even the smallest movement.

Types of Hair

In general, there are four basic types of hair: straight, wavy, curly, and kinky curly and these classifications are based on the shape of the hair fiber. Straight hair is the strongest and easily reflects light giving it a glossy appearance. Wavy hair has s-shaped curls down its length and tends to frizz and get split ends. Curly hair strand thickness varies from fine to coarse down the entire hair shaft. Kinky curly hair has the tightest curls, from fine to coarse, with s-shaped and z-shaped curls.

Texture

There are three basic types of hair texture: fine, medium, and thick or coarse. Texture describes the thickness of each individual strand and not how it feels. Typically, fine hair is thinner than a piece of thread, medium hair is the same thickness, and coarse hair strands are thicker than a piece of thread. The most fragile strands are fine hair. They can be easily damaged since fine hair has only two layers, a cortex and a cuticle. Also, fine hair is difficult to style. Medium hair is the most common and can easily be styled. Structurally, medium hair may contain a medulla and as such is stronger and does not break easily. Thick or coarse hair is stronger because it contains all three hair layers and therefore better at resisting strand breakage.

Hair Color

Natural hair color is due to the amount and type of melanin cells present in the hair follicle. Melanin cells produce two types of color pigments inside the hair follicle that are packed into granules in each hair strand. One pigment, eumelanin is the dominant pigment for black, brown, and dark-blond hair. The other pigment, pheomelanin or erythromelanin, is the dominant pigment for red hair. Little to no pigmentation in the hair strand results in blond hair. Hair turns gray when the melanocytes in the basal layer of the hair matrix are greatly reduced in number and the melanin pigment production ceases. In general, dark hair contains more melanosomes (pigment producing cells) while light hair contains fewer.

Hairstyles

Hairstyles vary widely depending upon the culture and the historic period. A person's beliefs, social position, and attitude are often expressed in their hairstyle (for example, dreadlocks). Healthy and youthful people have healthy hair. The color and texture of hair often indicates ethnic background. Facial hair is a sign of puberty in males, whereas white or gray hair is a sign of age as is male pattern baldness. Hair helps identify appearances and overall this is more important for females (gender identity) than males. Hair is the only body structure that is completely renewable without scarring. Religious practices also influence hairstyles, especially in women,

8. The Hairy Who Are Scary

in that various religions during history have had a direct effect on how hair is worn and covered.

Hair Loss

Alopecia is referred to as hair loss and in reference to the head is commonly called baldness (male pattern baldness in men). Some types of alopecia can be caused by an autoimmune disorder and extreme forms result in the total loss of all hair from the body. Other causes of baldness include fungal infections, trauma (such as compulsive hair pulling), radiotherapy or chemotherapy, and nutritional deficiencies such as iron. Another method to control hair loss is the use of testosterone suppressor drugs to keep the overproduction of hair growth in check.

Another form of losing hair from the skin surface is called depilation and is the removal of the entire hair strand by either trimming, plucking, or shaving and usually involves terminal hair and not vellus or lanugo hair. Most hair removal is done with shaving but waxing is also a popular way to remove hair. A haircut is considered the removal of a significant amount of hair whereas a trim is mostly removing split ends while leaving the overall shape and look of the hair unchanged. So cutting hair removes more than a trim. Cut hair appears thicker because the tapered split end has been removed but in reality the thickness of the hair strand is unchanged after cutting.

Hair Restoration

Most hair restoration entails the removal of hair from one part of the scalp and transplanting it somewhere else instead of the actual regeneration of new hair growth. A recent study, "Microenvironmental reprogramming by three-dimensional culture enables dermal papilla cells to induce de novo human hair-follicle growth," published in the *Proceedings of the National Academy of Sciences* has shed light on what it would take to generate new hair growth.

This method offers the possibility of inducing large numbers of hair follicles or rejuvenating existing hair follicles, starting with cells grown from just a few hundred donor hairs. It could make hair transplantation available to individuals with a limited number of follicles including those with female-pattern hair loss, scarring alopecia, and hair loss due to burns.

It is known that dermal papilla cells give rise to hair follicles so the cloning of hair follicles using inductive dermal papilla cells has been around for some time. The above study showed that when these cells are aggregated together they give rise to new hair cells but if left as individual cells they do not. The implication is that clumps of hair cells together will be more efficient in new hair growth so this is something we must keep in mind during our favorite hairy film moments.

Physiology

Some hair restoration procedures involve hair transplants from one area of the scalp to another. For this procedure it is important that enough donor hairs are available so some people such as those with baldness problems (and burn victims) most likely will not benefit. It should be noted that this transplant process does not create new hair and often does leave scars. These transplants contain dermal papilla cells for continued hair growth.

The Skinny on Skin

To properly discuss hair we must also talk about skin. After all, that is where the hair grows. As goes the skin, so goes the hair. Skin is the largest organ mammals have. For adult humans the skin weighs an average of 4 kg and covers an area of approximately two square meters. Its major function is to act as a barrier against the surrounding environment, the harsh and unforgiving outside world. Skin also serves as a shield to protect the delicate organs and tissues within our bodies. Skin is composed of three major layers of cells and tissues. The outermost layer is called the epidermis, and this serves three functions: it is a physical barrier, it protects against light (such as harmful radiation), and it is an immunologic organ (it helps fight off germs and bacteria). Subsequent layers, the dermis (or corium) and the subcutis, contain cells that perform specific functions that maintain the integrity and action of skin.

Each layer of skin has its own distinct specific cell types. The outermost layer, the epidermis, is composed of the basal cell layer, primarily consisting of keratinocytes, melanocytes, and merkel cells. The next layer is called the squamous cell layer and is composed of Langerhans cells and the desmosome-tonofilament complex. Then comes the granular cell layer, the horny cell layer, the stratum lucidum, and the oral mucosa. The epidermal appendages consist of eccrine glands, apocrine glands, hair follicles, arrectores pilorum, sebaceous glands, and nails. The basement membrane zone contains ultrastructural components and chemical components.

Importance of Sweat Glands

About 2.5 million years ago, during the evolution of man from ape, man began to lose body hair and at the same time developed sweat glands all over the skin thereby enabling humans to perspire over most of their body, which eliminated the need for excessive body hair, as well as eliminating the need for panting.

Fur-bearing mammals do not sweat (with the exception of horses) so they pant to help blow off excessive heat. Humans do not pant so they sweat to help cool down the body. Wolves pant whereas humans sweat. So, when a human transforms into a werewolf he most likely would not sweat but, rather, pant to cool down. Or perhaps a combination of the two would get us a mildly perspiring and tongue-wagging beast.

8. The Hairy Who Are Scary

A sweaty werewolf would have matted down hair that could be uncomfortable. The change in body hair could affect the change in the number and function of sweat glands and certainly how body thermoregulation occurs.

Adult humans have about 2 million sweat glands, those that cover the body are called eccrine glands whereas those in the armpits and groin are called apocrine glands. These glands produce watery fluids that help cool the body by evaporation thereby releasing heat. All of these glands likewise produce a fatty secretion that helps lubricate the hair shaft that also helps to keep it from drying out and becoming brittle.

Excessive Hair Growth

Though the exact reasons for excessive hair growth are currently unknown the underlying causes are hormonal and related to excessive testosterone, a male androgen hormone. It follows that since hair growth is hormonally related then hair removal or shedding must also be hormonally related.

In women hirsutism can be either congenital (present at birth) or acquired during life. Exposure of women to excessive male hormones such as testosterone is the primary cause of acquired hirsutism that may also include acne, a deepening of the voice, irregular menstrual cycles, and an overall more muscular body shape. Hirsutism occurs in about 10 percent of women between 18 to 45 years of age.

More common in men, when terminal hair grows in areas that would normally have vellus hair it is called hypertrichosis (also called Ambras syndrome; excessive cases have also been referred to as "werewolf syndrome"). The excess hair growth, both length and density, in hypertrichosis cases may consist of any type of hair (lanugo, vellus, or terminal). The first reported case was in the late 16th century and so far around 50 cases have been reported since then making this symptom rare indeed. Interestingly, those with hypertrichosis often have gingival (gum) disease and other dental problems.

Hypertrichosis hair is typically thick and pigmented and can be either generalized or localized on the body and either congenital or acquired. Those afflicted with this condition were often performers in sideshows at carnivals (for example, the bearded lady); many were considered freaks and promoted as having animal instincts. Congenital hypertrichosis, caused by a very rare genetic mutation, is inherited and present from birth; there is no known cure for this condition. Those with congenital hypertrichosis do not have any hair on glabrous tissues such as the palms of their hands, soles of their feet or on mucous membranes. This congenital form causes males to have excessive facial and upper body hair growth whereas females show less with hair distribution often being asymmetrical and non-uniform over the body. In addition to the generalized hypertrichosis there are also localized versions, one being "hairy elbow syndrome" in which excessive vellus hair grows on and around elbows. These localized areas have an increase in hair density and length.

Physiology

Hypertrichosis acquired after birth is usually the result of the side effects of some drugs, cancer, and possible links to certain eating disorders. This form is characterized by rapid growth of lanugo hair, especially on the face (cheeks, upper lip, and chin), with other areas being the trunk and armpits (palms and soles are not affected) that have less. Also, some forms have multiple hairs growing out of the same follicle. To treat acquired hypertrichosis the underlying cause must be removed, such as drugs, hormones ... or exposure to the rays of the full moon.

When hypertrichosis occurs areas of the skin transform the small vellus hair follicles into the larger terminal hair follicles. Though this normally occurs during adolescence, especially in males when underarm and groin vellus hair follicles transform into terminal hair, this switch can also occur in adults and involve areas of the skin that do not normally make terminal hair. When hypertrichosis occurs in a pattern formation on the skin, this is usually a sign of an internal malignancy, an irritation, or trauma. This hair is more dense and longer than normal terminal hair.

Hair removal is the easiest way to control hypertrichosis and can either be temporary or permanent. Temporary hair removal can last from a few hours to a few weeks, depending upon the method used. Some methods are cosmetic, such as trimming or shaving, where the hair shaft is cut at the skin level, whereas others, like electrology and waxing, remove the entire hair root and bulb and may last a few weeks. A popular method for hair removal is using laser technology, but this does not work on hairs without any pigments, such as white hair. The laser targets the melanin dye in hair follicle (lower ⅓ of hair shaft) and white hair does not have any melanin.

Hypotrichosis

The loss or reduction of hair is called hypotrichosis and is typically seen when vellus hair grows in areas of the body that normally have terminal hair. The vellus hair is usually fine, short, brittle, and often lacks pigmentation. This hair loss condition mostly occurs after birth when normal hair is shed and replaced by vellus hair and, for some males, baldness may result by the age of 25. Also, balding is when terminal hair switches to vellus hair.

Lycanthropy

Lycanthropy has its origins primarily in 16th century Europe, though it can be dated as far back as the ancient Romans. According to Roman mythology, the god Jupiter transformed the cruel, cannibalistic King Lycaon of Arcadia into a wolf (Lycos is Greek for wolf—and where we get the word lycanthropy from—and lycorexia means the craving for raw meat). Curiously, during the 16th century all reported cases of lycanthropy occurred primarily within the peasant class in low lying areas of Europe that were under 500 feet above sea level whereas regions above 500 feet have no reported cases. As a rule, peasants could not afford wheat bread, which was saved for the upper class, and therefore had to eat cheaper rye-based bread. Unfortunately, rye grains were contaminated with ergot fungus, and when the peasants harvested the grain the ergot fungus came along with it. (Ergot does not contaminate wheat so the upper class was never exposed to it.) Ergot contains natural hallucinogens (like LSD) that, when consumed in quantity with bread, induce psychotic episodes. The peasants who had a lot of the ergot fungus in their system were subsequently very susceptible to outside influences. And since werewolves were popular at the time, the power of suggestion convinced many of the peasants in the low laying regions that their communities were rampant with werewolves (therefore, much of our favorite werewolf lore originated through the use of mind-altering drugs).

The Films

The following baker's dozen films are good representative examples of the types of hair changes seen in some of our favorite SF films. The breakdown is nine films with werewolves, one hairy Hyde, one throwback, and two apes.

The following films range from the classic full-moon induced transformation, to biochemical induction and hormonal and glandular alterations. It should be noted that in just about all of these films our heroes do not have any moustaches in their hairy state. Hair seemly grows everywhere, even in places it wasn't meant to, but somehow hair on the upper lip has no role and is a no show in these films.

***Dr. Jekyll & Mr. Hyde* (1932)**

A pre-code horror film that earned Frederic March an Academy Award for Best Actor for enacting no less than five transformations from Jekyll to Hyde in this film. The first transformation results in simian looks with hair growth on Jekyll's brow, hairier eyebrows, and on the side of his face and back of his hands. Subsequent transformations show a Hyde with bushier head hair and more on the brow, face, and hands. The first two transformations into Hyde are induced by Jekyll's drinking a formulated concoction, whereas later transformations are spontaneous, implying a hormonal basis for the changes. Since the initial reversals back to Jekyll are spontaneous, it suggests the effects of his concoction are temporary.

It should be noted that some hair strands on Hyde were shorter than before the transformations from Jekyll. Also, as Jekyll some hair strands were wavy but as Hyde all head hair appears to be straight. The third transformation was spontaneous resulting in thicker forehead hair, longer strands of his facial hair (long vellus hair), and even bushier eyebrows. Hyde reverts back to Jekyll after drinking another concoction (same as before or perhaps some sort of antidote). Hyde gets progressively hairier with more pronounced features during each successive change. After the fifth transformation Hyde is shot and he reverts back to Jekyll while dying.

Of note is that there were no moustaches during any of the transformations. So, some vellus hair, on the brow, face, and hands, was able to convert to terminal under a chemically induced hormonal burst that reverted after some time. Also, wavy and curly hair reverted to straight hair and back again. Ultimately, it was mental anguish that catalyzed the subsequent hormone signals that resulted in the hair transformations.

The Werewolf of London (1935)

Though we see two werewolves in this film they appear to be essentially alike. The first, Dr. Yogami, who as a werewolf has a widow's peak brow, long bushy sideburns, bushy eyebrows, as well as hairy hands. A key plot device is the plant *Mariphasa lumina lupina* and stem fluid from a blossom is what is needed; apparently not much is required to (ahem) stem the change to werewolf.

In researching werewolves Dr. Glendon (Henry Hull) opens a book and reads an entry, *"De Lycanthrophobia (transvection from man to werewolf)"*: "The essence of the Mariphasa blossom *squeezed into* the wrist through the thorn at the base of the stem is the only preventive known to man. Unless this rare flower is used the werewolf must kill at least one human being each night of the full moon or become permanently afflicted."

The fluid must break the skin surface and enter the blood stream, muscle tissue, skin tissue, as well as localized hair bulbs and follicles. The immediate effect is a cessation of hair growth, which halts the transvection into a werewolf. Since it appears that just a small amount of the fluid enters the broken skin it must be very potent to counteract the hormone induced terminal hair transformations. To work, the stem juice is essentially a hormone-suppressing substance that changes hair growth from anagen to catagen. As Dr. Yogami (Warner Oland) notes, "it's not a cure, it's an antidote." The fluid from one blossom of each night of every full moon is needed.

Dr. Glendon's original transformation into werewolf was filmed with rapid cuts of the doctor walking behind columns with each cut showing a hairier and hairier beast. The terminal hair growth was selective and most prominent on his hands and face. In particular a widow's peak brow, very long sideburns, bushy eyebrows, hair on his lower lip (called a mouche or "soul patch"; note: no moustache visible) and hairy hands are the obvious changes. How much hair is on his chest, back, and legs is unrevealed, but based on his tight fitting clothes not much, if any, hair grew on these areas. Overall, Dr. Glendon shows selective terminal hair growth.

There are a total of five transformation (or transvection) scenes of Dr. Glendon in this film. The first is when he exposes his hand under the moon ray lamp and hair begins to grow on the back of his hand. For the second transformation we see the hands transform first, followed by an extended widow's peak of head hair on his brow, sideburns then lengthen and thicken, and the eyebrows also lengthen and thicken.

8. The Hairy Who Are Scary

For the third transformation we see the head change first then the hands. For the fourth transformation Glendon is at a secluded place and while laying down to rest the moon's rays come through a window transforming him. The fifth and final transformation occurs while Glendon and Yogami are fighting in the lab. The stress of the fighting in addition to the rays of the moon lamp cause the transformation. Dr. Glendon is shot as a werewolf and reverts back to human form as he dies and all the hair essentially vanishes.

Frankenstein Meets the Wolf Man (1943)

For many Baby Boomers and Monster Kids *The Wolf Man* (1941) starring Lon Chaney, Jr., is the definitive werewolf movie. However, the actual transformation scenes were minimal with a reverse transformation of Wolf Man back to Talbot being the only real filmed effect. In the Universal Studio film series the classic transformation scenes were magically created by makeup master Jack Pierce who set the standard of excellence from which all other werewolf films are judged. And of all the Chaney werewolf films, the best transformation from Larry Talbot to Wolf Man is found in the film *Frankenstein Meets the Wolf Man*.

The first view of the Wolf Man in *Frankenstein Meets the Wolf Man* is during the opening scene of the film where two grave robbers disturb Talbot's grave and after opening the coffin the rays of the full moon cause the transformation. At this stage all we see is the fully converted Wolf Man. For the second transformation, Talbot is lying on a hospital bed and we see his sideburns lengthen and hair appear on his lower chin. His eyebrows grow and blend into hair growing down to the bridge of his nose. Though there is no moustache or mouche there does appear to be a bit of hair on the outer area of his upper lip. Overall, as Talbot his hair is straight whereas as the Wolf Man his hair is a curly to a kinky curly variety. Also, as Talbot his hair is a uniform medium brown color whereas as the Wolf Man he has some gray hairs suggesting pigment loss in some hair fibers. While in the hospital Talbot is wearing pajamas, but after he transforms and is out on the street he is wearing different clothes so he must have changed to street clothes just after his transformation to a werewolf. As the Wolf Man, Larry Talbot's legs and feet are covered with terminal hair, but as the fully transformed Wolf Man his hair is cropped and trimmed.

For his third transformation the sequence of hair changes appear to be the same as before. We later see him back to a normal Talbot so for this to happen would require some significant hair follicle activity in reverse for terminal hair to change to vellus hair not to mention the shortening of some terminal hair, the unkinking other hair shafts, and making new pigments to change gray hair to brown hair color.

The final transformation occurs while Talbot is strapped to a lab table and our first view of this is in mid-transformation so there is hair on his face, brow, and chin that all grows into the full-furred Wolf Man for his smack down battle with Frankenstein's monster.

Physiology

Undying Monster (1942)

This is a case where mental angst prompted the transformation to werewolf and there was no organic catalyst. A strong mental trigger caused Oliver Hammond (John Howard) to imagine himself into a werewolf. This is not as unreasonable as it may sound. Emotions are hormone driven, so with the proper psychological influence many hormones can be produced in response that can in turn make physical changes to the body such as the conversion of vellus hair to terminal hair.

Our first brief glimpse of the werewolf is at 56 min into the film. Shown only in shadow form it appears to have wispy hair with bushy head hair, beard, and bushy hand hair. The second view reveals Hammond with his entire face covered with short, close-cropped, thick, and furry hair on his face as well as long, wispy hair on his hands. Amusingly, actor Hammond's moustache is still visible under all the fur. After he is shot dead he reverts back to human form. In an amusing bit of cinema "proof" a spectra analysis performed on a hair sample from Hammond confirms that it is identical to wolf hair. Quite a remarkable observation and conclusion.

The Ape Man (1943)

Dr. James Brewster (Bela Lugosi), gland expert develops a formula based on ape fluid that he tries on himself. The fluid causes permanent changes, resulting in terminal hair growth on Brewster resembling that of a chimpanzee. His colleague, Dr. George Randall (Henry Hall) comments, "We made the experiment and unfortunately it was a great success," adding a bit of macabre humor. (In my own research career I have never felt unfortunate over any successful experiment.) Brewster now sports hair down his brow, up his cheeks, on his chest (seen through an open shirt), and on his forearms with no moustache, no mouche, and no hair on the back of his hands. Overall, the hair growth is confined, localized, and straight.

Desperate for an antidote, Brewster says, "Human spinal fluid injected into me. The only way to counteract the ape fluid injection … (the spinal serum will) cure completely and permanently." Randall warns, "(The fluid) must be taken from a living person. And the taking means instant death" (actually it doesn't). In the film both the removal and injection sites are the lower spine. The effects of the first injection of human spinal fluid is temporary and soon wears off ("won't keep," says Brewster), prompting Brewster to kill several people to obtain spinal fluid. The container of fluid shown appears to have around 300ml., "half a dozen injections" says Brewster, (300ml at six injections in 50ml per injection, too large a volume to inject directly into the spine. Such an injection could cause severe pain and perhaps even a partial paralysis.). Throughout the entire film the excess terminal hair never goes away, suggesting the ape fluid caused a permanent, genetic-based reaction rather than a temporary hormonal response. Therefore, the hormonal effects of the serum caused a permanent though selective change in Brewster's vellus to terminal hair.

8. The Hairy Who Are Scary

Captive Wild Woman (1943)

Dr. Sigmund Walters (John Carradine), a famous endocrinologist, uses gland extracts to change a gorilla to a human. As Walters says, "Glands can transform physical matter into any size, shape, or appearance." To demonstrate to the audience that Walters knows his stuff we get a glimpse of a (fictitious) scientific journal, *Medicine and Surgery* (published monthly) with the special feature shown on the cover table of contents: "'GLANDS and Their Secretions' by Dr. Sigmund Walters (endocrinologist)." While reading some of the article we learn that Walters invented (fictitious) vitamin E2 that "determines the physical characteristics of all forms of animal lift" and that Walters also made "three attempts at racial improvements." Walters' goal is the "transformation of an animal into a human being." As such, there is quite a bit of hair activity from ape to human and back and forth.

A female patient presents to Walters who is diagnosed with a "follicular cyst which induces the secretion of an unusual amount of sex hormones" (read: estrogens). Seizing the opportunity Walters injects her glands (ovaries?) into the gorilla, Cheela, and we quickly see her hand lose hair (shed? absorbed?) and become human like. Therefore, Cheela's terminal hair was transformed into vellus hair. Once completely physically transformed to a human, now known as Paula Dupree (played by Aquanetta), though hairless, she still has her animal instincts. In one instance a jealously induced hormonal rage changed Paula back to Cheela suggesting an emotional/hormonal link to the transformations. For the transformation the skin first darkens on her face, hands, chin, neck, and midriff followed by slowly growing terminal hair (this growth rate seems more reasonable). In the transformation we see long wispy hair (long vellus hair?) on her legs and feet. Her eyebrows are also different. Her final looks with her hair combed back and facial hair makes her resemble Zira from *Planet of the Apes* (in reality it's the other way around).

Most hormone-induced effects are temporary and to support this Walters says, "Terrific emotion would destroy the new gland growths," so such an outburst would cause a reversion of the effects. For the transformations, Walters, in addition to the gland extracts (i.e., implantation of ovaries), transfuses about 50cc of the woman's blood into Cheela. Then later another 80cc of whole blood is transfused into Cheela/Paula. To sustain the human effects Walters has "grafted glands from a living woman into your body." As the hormonal effects wear off, the hair on Cheela's hands disappears (recedes?). Though male hormones like testosterone are responsible for hair growth, in the case of Cheela/Paula, the hormones are estrogen based so there must be a different hair follicle response for her. One interpretation is testosterone changes vellus hair to terminal hair whereas estrogen changes terminal hair to vellus hair.

The Werewolf (1956)

Duncan Marsh (Steve Ritch) is in an auto accident and falls into the care of two scientists Emory Forrest (S. John Launer) and Morgan Chambers (George Lynn).

Physiology

Chambers "treats" Marsh with an experimental serum, a "full inoculation" obtained from a wolf mutant who died of radiation poisoning. Unsurprisingly the radiated wolf blood turns Marsh into a werewolf. It is radiated mutated wolf blood that causes Marsh's transformations so as a result some of the wolf's radiation-mutated DNA was incorporated into Marsh's DNA thereby causing the selective transformations.

After realizing the horrendous results of the experiment, Forrest questions Chambers accusingly, "What have you done?" He replies, nonplussed, "Done? Accomplished is a better word ... that wolf man is the proof. Radiation creates mutants." The idea that radiation fallout from nuclear bombs would create mutants was a popular concept in the mid–1950s. Chambers aim is to immunize humans by a slow series of inoculations to protect against radiation poisoning.

Since Marsh's lycanthropy is the result of science gone wrong, it may explain why his transformations into werewolf and everything related to them are random. Transitions often occur during daylight, and are incomplete, with clumps of hair appearing on areas such as his face, hands and wrists rather than all over his body.

While in werewolf form Marsh's left foot is caught in a bear trap (similar to Larry Talbot in *The Wolf Man*) and his trouser leg ripped, showing no hair on his shin and lower leg, again showing selective terminal hair growth. The hair seen is relatively sparse and not especially thick, unlike the Talbot werewolf. At one time the Marsh werewolf has hairy feet, which are not so hairy other times.

Marsh also has a wedding ring that comes and goes while he is in werewolf form. For his first transformation we see his hairy hands bearing the ring, but later his transformed hand is seen without the ring. Furthermore Marsh's signature five o'clock shadow comes and goes with transformations. Unique in this genre is that the Marsh werewolf sports a handlebar moustache that is long and curved along his cheek.

Marsh's emotions are at least one catalyst to his becoming a werewolf. His first transformation occurs during a fight. Later, when he faces the threat of being shot, we see facial hair appear on his cheeks and eyebrows, his hairline descend into a widow's peak, progressively hairy ears, nose, and the reappearance of his moustache. To further flout convention Marsh also transforms into the werewolf once while sleeping in a jail cell. Inevitably, after being shot to death the werewolf transforms back to Marsh, but in death he is clean shaven with no five o'clock shadow.

Monster on the Campus (1958)

Professor Donald Blake (Arthur Franz) discovers that primitive DNA from the coelacanth fish will revert a more advanced species into a more primitive form, in essence, create a "throwback." He witnesses an exposed German Shepherd transform into a wolf-like creature with elongated teeth and a mean disposition. After Professor Blake accidentally scrapes his hand against the sharp coelacanth teeth, he immerses the wound in fluid from the same container the fish was transported in, flushing the

8. The Hairy Who Are Scary

fish DNA directly into his bloodstream. He becomes ill and gradually transforms into a violent prehistoric man.

The first transformations were not rapid and took some time to take effect. For the scene at the cabin the professor changes within seconds into the monster after a self-injection of coelacanth serum. The effective transformation scene of the professor turning into the throwback Neanderthal in the cabin is done with lab-dissolves showing a hairier and hairier head. Vellus hair turns into terminal hair on the face, brow, chest, and hand, though no moustache hair is seen. The hairier throwback Neanderthal was chosen to suggest more primitive characteristics and more hair was effective in achieving this goal.

Apparently, it does not take much to catalyze a transformation since in one scene some coelacanth blood dripped into the professor's pipe and after inhaling some smoke of the burning serum he transformed into the throwback. (It should be noted that the heat of burning of the serum in a pipe does not alter the serum's effects; heat would denature most biomolecules making them useless.) Since the serum effects came via pipe smoke it can then be considered an inhalant and therefore must be absorbed through the lungs. After being shot dead the professor reverts back to human form and the hair is absorbed/recedes back into the body.

The Curse of the Werewolf (1961)

What separates this werewolf from all the others is the presence of hair on the glabrous skin of his hands. Early in the film a young adolescent Leon clearly shows growth of hair on the glabrous skin of the palm of his hand and later, as an adult, we see another clear view of hair growing on the palms and fingers of his hands.

In werewolf form his eyebrows disappear and excessive hair grows on his chest, hands, arms, and his sideburns lengthen. No moustache hair is visible. It is unknown about hair on his legs and feet as an adult werewolf since he wore tight pants and boots.

As a young adult it took a lengthy exposure to the full moon to cause the transformation of Leon (in other werewolf films the transformation occurs as soon as the

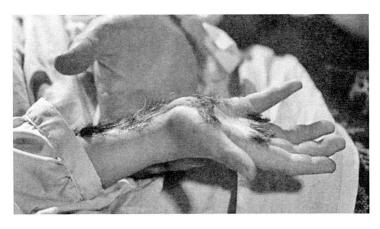

The Curse of the Werewolf provides the only example in werewolf cinema where hair is seen growing on glabrous skin. Note that the back of his hand does not have any hair growth.

107

full moon is seen). Perhaps an adolescent prepubescent Leon did not have enough testosterone in his body to make an effective transformation to change all the vellus hair to terminal hair. Like Larry Talbot in *The Wolf Man*, we see Leon's legs/feet change first and, just like Larry Talbot, Leon then changes pants and puts on a shirt after the transformation.

Though slow to start, once started, the full transformation rapidly occurs. Leon has sideburns as an adult that remain so after reverting back to human form suggesting selective vellus to terminal hair transformations. For the last transformation hair first grows on Leon's glabrous skin on the palms of his hands then as patches on his back and chest. Then his eyebrows disappear and finally hair grows on his face.

The Beast Must Die (1974)

Though this film is more of a detective mystery to guess "Who is the werewolf?" than a true horror film we nevertheless see a transformation. For most of the film the werewolf seen is actually a real wolf so, as such, the hair transformations would be different from the more traditional human-wolf hybrids.

The first (and last) werewolf seen is Jan. Jan bites Caroline Newcliffe who later becomes a werewolf. During her transformation her hands change first and become hirsute hairier (looks like terminal hair and not excessive vellus hair). Then she is instantly in wolf form. As a wolf her hair is shaggy, puffed up, and not matted down as normal, suggesting her piloerector muscles are fully tensed. As a man Jan has a moustache and as a wolf there is no hair on the upper wolf snout. The only actual transformation we see is a reverse one of Jan as a werewolf, after being shot dead, reverting back to his human form (thereby solving the "mystery"). The hair recedes and his moustache reappears demonstrating remarkably selective hair activity.

Werewolf Woman (1976)

This film features a blond female, Daniela Neseri (Annik Borel), who is in a trance and while dancing under a full moon, transforms into the werewolf woman. As the hirsute horror the woman has extensive blond hair all over her body (chest, back, legs, arms) that appears to be thin like vellus hair and not thicker terminal hair. Also, she has no hair on her cheeks, brow, and chin. In addition to her werewolf body hair she has dense unibrow hair that merges down on the bridge of her nose. Remarkably, her blond head hair is unchanged (long and wavy with "body") and her glabrous skin has no hair. Not surprisingly, as a werewolf she has no moustache, no mouche, nor hair on her under forearm area from the wrist to the underarm though hair is on top of the forearm and upper arm, though not as thick as on her torso and legs. Also, as a werewolf she has no underarm hair as well as no hair on the bottom of her feet. With all that blond hair she therefore has no pigmentation in the shafts of each hair strand cortex.

8. The Hairy Who Are Scary

The Howling (1981)

This film features a colony of werewolves from young to old, both male and female (it is amusing that old werewolves complain of teeth problems). The male werewolves in the film have long, thick terminal hair all over their bodies whereas the females have wispy hair. The movie differs from others in the genre in that emotions seem to catalyze transformations more than exposure to the full moon, and the werewolf form itself fluctuates. In one scene, Eddie (Robert Picardo) transforms during the day, inside a house, and into a form that is neither wolf nor human. His hair becomes long, thick, and wispy and doesn't evolve into dense wolf fur.

When Bill Neill (Christopher Stone) transforms into a werewolf he develops long wispy hair on his neck and body in patches. As Neill he does have a small, well trimmed moustache that more or less stays the same when he is a werewolf. The female he is with, Marsha (Elisabeth Brooks), also shows long wispy hair on her face, neck, arms, and upper chest. She has longer hair on her forearms and the backs of her hands.

Karen White (Dee Wallace) is also indoors when she transforms into a werewolf at end of the film. She has long blondish hair all over her face including her upper and lower lip, though her final appearance is more feline than wolfen.

An American Werewolf in London (1981)

The highlight of this popular film is the technical tour de force of the transformation scene. You feel the pain of the vellus hair changing into terminal hair as the density of the hair increases and forces its way through skin. David Kessler (David Naughton) has long wispy thick strands of hair that first appear on the back of his hand, then down the length of his spine, emerging on his forearms, then upper chest, neck, and lower legs. Remarkably, the hairs lengthen visibly and become thick and furry as they cover his body. The hair on his head becomes shaggier and facial hair is the last to appear with modest sideburns. All the early hair growth is black and after much of it filled in some of it was gray indicating modest pigment lost. After David is shot and killed he reverts back to human form, with the excess hair being absorbed or receded.

Summary

Human bodies are covered with many types of hair. Though the hair root is essentially the same the hair follicle that grows out of it can vary depending upon a number of factors, both genetic and those acquired by life styles.

For the werewolf films, how could the rays of a full moon bring about the change of vellus hair to terminal hair? One possible explanation is that the moon's albedo

Physiology

slightly alters the wavelength of light that in turn excites in some people the overproduction of hormones, especially testosterone, which could induce excessive hair growth. Once the full moon is gone the person reverts back to normal since the stimulus is no longer present.

When contemplating the werewolf, excessive hair growth is relatively easy to understand but what is difficult to understand is how hair could retract and be reabsorbed instead of shed. Such reabsorption would require radical biochemical and metabolic processes that give off heat thereby making the skin surface warm to the touch.

Hair can come and go and either grows from where it belongs or grows anew from places it wasn't meant to grow. In reality, instead of being reabsorbed, hair is simply shed and the bulb regrows a new hair follicle. When you next see your favorite hirsute hypertrichosis transformation scene you can better appreciate what's going on, both under and on the skin.

9

Drugs in Science Fiction Cinema

Introduction

Drugs became prominent in the public eye when penicillin was discovered in 1928. At the time it was considered a "miracle drug." During the counter-culture revolution of the late 1960s drugs became much more of an issue to the general public and the meaning of the word, drug, changed from something inherently useful to something that has bad connotations. No matter since drugs, good, bad, and sometimes useful, are here to stay. Though drugs are often used in SF films it is the fictitious ones that are of interest.

[Disclaimer: Do not operate heavy equipment while reading this section]

Better Living Through Chemistry

In total, there are about 10,000 approved pharmaceutical drugs and these can be sorted into about 50 categories. Some categories have many hundreds of drugs, so the options are extensive. Those categories with the most drugs available are cardiovascular (heart), antibiotic (infectious), and antineoplastic (cancer). In general there are two major types of drugs, either synthetic, meaning they are chemically made, or natural, meaning the ingredients are obtained from nature. Those drugs made from natural products such as plants and microorganisms are perceived differently than those synthetically manufactured.

SF cinema drugs come in all shapes and sizes, flavors and colors, bitters and sweets, ingestible, inhalable, or injectable. And their power to work is extraordinary. For cinema drugs they are not the recreational kind nor what we would all consider standard like aspirin, insulin, or blood pressure medications, but rather, those fictitious drugs that are used as conveyances or shortcuts to advance plots. We will be taking a closer look at some of these drugs and perhaps learn if there is any basis in reality in how they work. In the pharmaceutical industry how drugs work is called "mechanism of action" or MOA. The MOAs of cinema drugs, all things considered, are really not that far removed from how real drugs actually work. As it turns out, many of the cinema drugs, though they may have fancy names, are essentially ordinary drugs with ordinary MOAs.

Physiology

The word "drug" is believed to originate from an old French word, "drogue," meaning "dry barrels" a reference to containers with medicinal plants preserved in them. A drug is any substance that may have medicinal, intoxicating, or performance enhancing effects.

Foods or food-related product are generally not considered drugs. That being said, what one culture considers a drug another culture may define as a food, so the distinction can blur. Governments also often play a role in determining what is or is not a drug.

Drugs are powerful substances that can do harm if not used correctly, so in many countries they are regulated by law and dispensed only through prescriptions. Since many of the approved 10,000 drugs are available only by prescription this may or may not pose a problem for our annoyed cinemascientist. If he has a medical degree then he can easily obtain the drugs of choice himself by writing his own prescription, a perfectly legal procedure using his own medical credentials. Otherwise, he would either have to obtain the drugs from a colleague (willingly or otherwise), steal them, obtain them from the black market, overseas where less questions are asked as long as the money is good, or perhaps make it himself. And, of course, he will need a large enough supply of the drug(s) to serve his needs. It should be noted that more often than not his drug of choice is so rare or experimental that no pharmacy carries it so he must make it himself.

Many of these drugs have special problems when combined with certain other drugs and collectively these are called counter indications. One drug can inhibit the MOA of another drug thereby neutralizing the effects making the drug(s) essentially useless or perhaps enhance the effects making it potentially life threatening. It is the combinations that are the issue and not the individual drugs in that one drug can up regulate a biological process while another down regulates and when combined could result in serious physiological trauma, coma, or even death.

The definition of a pharmaceutical drug is any chemical substance that is used in the treatment, cure, prevention, or diagnosis of a disease. These drugs can also be used to enhance physical or mental well-being. Such drugs can be taken for a limited time or on a regular basis, depending upon many factors such as the duration of the illness and the overall health of the patient.

Another class of drugs are the recreational drugs which are primarily psychoactive substances designed to enhance pleasurable experience. They are also chemical substances that affect the central nervous system. (The central nervous system, separate from the peripheral nervous system, contains the majority of the nervous system, including the brain, spinal cord, retina and cranial nerves.) As such, some of these drugs can be addictive with habitual use. Many natural substances such as beer, wine, and certain mushrooms travel the line between food and recreational drug and when ingested do indeed affect both the body and mind.

When considering recreational drugs marijuana is often the first thing to come to mind. However, the most common recreational drugs used in the world are alcohol and tobacco along with betel nut and caffeine. Other recreational drugs are ayahuasca,

9. Drugs in Science Fiction Cinema

a psychoactive drug mostly used by shamans in the Peruvian Amazon; peyote, primarily used by native Indians in Meso-America and South America; opium, primarily used in Asia; and Khat, an amphetamine-like stimulant used in Islamic countries. Some hallucinogenic drugs are LSD, mescaline, peyote, psilocybin, and ecstasy.

The cultivation, use, and trade of drugs, in particular psychoactive drugs, began before civilization. Psychoactive drugs that are used to invoke altered states of awareness for spiritual or religious purposes are called entheogens and are regularly used in ceremonies and rituals. Entheogens are mostly hallucinogens that are either psychedelics or deliriants but some are also stimulants and sedatives.

Drugs That Medicate

Pharmaceutical drugs have a medicinal purpose and aim to cure or to alleviate particular symptoms of an illness or a medical condition. These drugs can be obtained by prescription, or over-the-counter, or behind-the-counter in which a registered pharmacist dispenses the drug without needing a physician's prescription. These regulations vary from country to country and culture to culture.

Most non-generic drugs currently available have been patented meaning the developer of the drug has exclusive rights to produce and commercialize it. This also prevents others from making and selling the patented drug. Those drugs that are not patented are called generic drugs because they can be produced and sold by anyone. At one time most of these generic drugs were patented but the patent has expired (patents have a life of 17 years of exclusivity and after the patent expires the drug then becomes a generic).

Since the dawn of man the spiritual and religious use of drugs has been very common. The cultivation, use, and trade of drugs, in particular psychoactive drugs, began before civilization. Drugs that have spiritual and religious use are called entheogens and some religions are completely based on the use of these drugs. Entheogens are mostly hallucinogens that are either psychedelics or deliriants but some are also stimulants and sedatives.

In the United States such regulations are determined by the Food and Drug Administration (FDA). As a regulatory body the FDA was brought into being by President Theodore Roosevelt in 1906 to ensure drugs were made safely and used properly. However, it wasn't until 1938 that the government gave the FDA authority to monitor the safety of new drugs. World War II put everything on hold and after the war the era of modern medicine began with the FDA playing a larger role in drug policy.

Smart Drugs

Nootropics (or smart drugs) are designed to improve mental capacity for memory, concentration and learning. These drugs have been used to treat such conditions as attention-deficit disorder (ADD), Parkinson's, and Alzheimer's. They are also prescribed for patients who have lost some brain function as a result of aging.

Stimulants

Amphetamine (speed), cocaine, Dexedrine, methamphetamine, and Ritalin are examples of stimulant drugs, or "uppers." Caffeine as well as theophylline (the main active ingredient in tea) are also stimulants. Stimulants enhance the neural activity of the central and peripheral nervous system and induce temporary improvements in mental and physical functions, including increased alertness and focus. Heart rates and blood pressure increase with stimulant use.

Administration of Drugs

Since not all drugs are the same it follows that not all drugs are taken in the same way. For example, drugs given as an injection can be intramuscular, intravenous, intraperitoneal, or intraosseous; some drugs are inhaled as an aerosol; insufflation drugs are snorted through the nose; drugs taken orally must be absorbed by the intestines; suppository drugs are those given rectally which are absorbed directly by the rectum or colon; sublingual drugs are those placed under the tongue that diffuse into the blood stream; and topical drugs are those used as a cream or ointment that are absorbed into the body.

Drug Addiction

The nature of addiction and why addicts are so desperate is beyond the scope of this article. Suffice it to say much research and effort is going into understanding the causes and ways and means to treat the disease, which indeed it is. However, for you gentle readers there is a brain hormone called dopamine that may be a significant player in drug addictions, though some neuroscientists are debating this.

Dopamine is a natural neuromodulator that is associated with reward and is the engine that powers attention to goals, the need for urgency, and whatever it takes to get there. Many addicts have a dopamine spike when they get high on drugs and this only helps the reward process in wanting more and more again and again. Neuroscientists have been able to separate the neural circuitry that functions with liking and wanting. The liking element is mediated by the body's own opioid-like molecules where the wanting element is mediated by dopamine, the pleasure and therefore the pursuit of that pleasure. Though not obvious, key elements of addictions are the emotional and personality makeup of the addict, or as often in our case, the cinemascientist.

9. Drugs in Science Fiction Cinema

In SF cinema the drive of "it would be great to have some now!" and "how can I get some now?," though not particularly obvious, are the reasons for behavior justification in getting the wanted drug. Plus the drugs are used as a means to drive plots. Also, there is the ability to make one drug look like another though have an entirely different effect and MOA. In the real world, drugs have a limited MOA that does not apply to fictitious SF drugs that can work in any way intended to advance the plot. In the world of drugs in SF cinema the possibilities are endless.

Natural Products vs. Synthetic Drugs

Natural drugs are small molecules isolated from plants, fungi, bacteria, and other microorganisms and used to alleviate and treat diseases. Such natural products have mostly been the basis behind many of the FDA approved drugs and they still inspire drug hunters and makers to search for more effective treatments. Most drugs start out as a plant extract which is then analyzed to determine the active ingredients. When possible, these components are synthetically manufactured to bring down the relative costs of manufacturing.

It is generally cheaper to synthetically make a drug. For example, it is much less expensive to make aspirin than it is obtain through a plant source. But some natural drugs are made up of complex molecules that are difficult to replicate, so they are harvested from natural sources. Traditional medicines have used local flora and fauna for centuries indicating that they do have a basis of effectiveness. The general cost of natural product drugs is typically more than their synthetic counterparts due primarily to all the labor intensive steps involved compared to the cheaper synthetic steps.

Placebos

The word placebo is derived from the Latin meaning "I shall please," and refers to any simulated medical treatment that is based on deception. A common name for a placebo is a "sugar pill." In medical research placebos are typically given as control treatments as a way to separately measure and compare the effects of the actual study drug. What is interesting is that some patients given a placebo pill have perceived or actual improvements in their conditions which is known as the "placebo effect." What this means is the brain's perception and the brain's role in physical health is important. It should also be noted that placebos work differently in different diseases.

Non-drugs

To better understand drugs, it is important to also understand what substances are typically not classified as drugs. Natural foods aren't considered drugs, even though certain chemicals derived from natural foods do qualify. Vitamins and neutraceuticals are not considered drugs since they relate to nutrition and supplementation and do not have the Method of Action (MOA) of classical drugs. That being said, foods and neutriceuticals can help regulate blood pressure and fight infection. Poisons and toxins are also generally not considered drugs; however, purified extracts of poisons and toxins, are considered drugs. A drug, natural or man-made, must be introduced into the body. Hormones like insulin are natural biochemicals and are not generally considered drugs. However, for diabetics, insulin is classified as a drug because they need to inject the substance.

Regulating Usage

Many drugs have been prohibited or banned through government or religious laws.

The most common groups of banned drugs are barbiturates (Nembutal, Seconal, Amytal), bendodiazepines (Restoril, Normison Flunitrazepam, and Erimin), and opiates (Morphine, Codeine, Oxycodone, Heroin). Drugs like alcohol and tobacco are legal to use, but certain measures have been put in place to control usage, such as age restrictions and label warnings, not to mention a "sin tax," in which the consumer pays a premium to buy the product.

Drug use and abuse are major concerns of society as a whole and much research is being done on the causes of addiction and ways to treat the disease. Dopamine is a neurotransmitter associated with reward that may be a significant factor in addiction, though some neuroscientists are debating this. Many addicts experience a dopamine spike when they use and this reinforces the process.

Though the intention of drug banning is to prevent the use and abuse of potentially dangerous drugs, theoretically the most effective way to regulate use is through the prescription system, where the manufacture, distribution, marketing, and sale of drugs can be tightly regulated. However, this does not concern our cinemascientist who, either through illegal trade, the black market, or kitchen chemistry, can access whatever suits his SF fancy.

The Films

When looking at the overall list of drugs used in some of our favorite SF films some generalizations can be made. In these film most of the drugs work by altering brain function, directly or indirectly, usually in some sort of hallucinogenic way (perhaps as behavior controls?). Other drugs work like amphetamines or have hormone-like effects in that they influence metabolism and control cell growth. And, of course, they all do *exactly* as the drug's inventor intended. As a safety precaution perhaps the drugs in SF films should have warning labels not to mention expiration dates.

Science fiction movies made before 1960 featuring drugs as part of the plot are

9. Drugs in Science Fiction Cinema

relatively scarce, with only four included here. After 1960 drugs became much more prevalent in SF films. One of the reasons for this is that the availability of drugs, at least the synthetic kind, is a relatively new phenomenon. Pharmaceutical drugs weren't widely available prior to World War II and this relative absence is reflected in film. When drugs became more prominent in society, post–World War II and into the advent of "modern medicine," after 1960, this was also apparent in the movies.

To keep it simple, we focus on twenty movies. This is not an exhaustive listing of all SF films that feature drugs, real or otherwise, as part of the plot. That would be a long list indeed. To get a sense of the extensive possibilities, look no further than any Philip K. Dick novel.

The Invisible Man (1933): Monocane

Monocane is "a drug made from a flower grown in India. It draws color from everything it touches," according to Dr. Crawley (William Harrigan), the employer of Dr. Jack Griffin (Claude Rains). In reality, for anything to be rendered invisible, the physical optical density of matter would have to be changed.

To extract and test a substance like monocane, a compound from a plant and therefore a natural product, would require some sophisticated lab apparatus, mostly glassware to isolate the natural product's main ingredient in the plant.

Here invisibility results following Griffin's self-administration of Monocane, "a little bit of this injected under the arm every day for a month." It is unclear what a "little bit" is. Furthermore, it is unknown if the effects of monocane were a gradual progression to invisibility or an immediate response. While describing the injection he holds a large open glass container with approximately 500ml of a dark liquid, presumably Monocane. And an injection under the arm (in the arm pit?) is unusual; the actual injection site would more likely would be the bend of the arm due to easy access or possibly in muscular areas like the deltoids or buttocks.

The Invisible Man Returns (1940): Duocane

In *The Invisible Man Returns*, the word Monocane is not used. Instead, the name "Duocane" is used and described as "an extract of an east Indian herb that took the color out of things … combined with other ingredients."

To extract the active ingredient, duocane, from a plant a large amount of exotic glassware would be needed to do all the purification steps and this glassware is seen all scattered around the lab benches. Since duocane is derived from a plant source it can be considered a "natural product," meaning it is not synthetically made, and therefore one that could be regulated differently by the drug governing bodies like the FDA. Overall, there are more stringent requirements for synthetically made drugs than for naturally obtained drugs. Natural drugs are extracted and purified whereas synthetic drugs are manufactured so their FDA requirements are different.

Physiology

The Amazing Colossal Man (1957): Sulfa Hydral

Sulfa drugs were first discovered in 1935 and consist of a family of antibiotic drugs that are used to treat bacterial and some fungal infections. Sulfa drugs work by interfering with bacteria and fungi metabolism and were considered "wonder drugs" before penicillin was developed. Sulfa drugs concentrate in the urine before being excreted which is why they are primarily used today to treat urinary tract infections.

Here is an interesting verbal exchange between Dr. Linstrom and his colleague Dr. Coulter in the film. After a long night of research Dr. Coulter says, "the answer is in the bone marrow. We were so close we couldn't see it." Dr. Linstrom immediately senses what he was saying and responds with, "inject sulfa hydral compounds into the bone marrow." [Note his use of the plural of that word.] Though there are a number of sulfa drugs available none of them are named "sulfa hydral" so it is anyone's guess as to what that is and what it does to bone marrow. Dr. Coulter then says, "the thing that fooled us was we were looking for some unknown quantity in the plutonium radiation while all the time it was acting to a degree the same as a hydrogen exposure [whatever that means]. The secret was in the degree of exposure." Dr. Linstrom replies with, "then the injection of the sulfa hydral compounds should correct the body's regenerative balance ... it may stop his growth but it won't diminish his size." This is in reference to Glen Manning, the Amazing Colossal Man (see the article, "Amazing Colossal Science" elsewhere in this volume). Dr. Coulter added, "the stimulation of the hormone secretions in the pituitary or growth controlling glands will take care of that ... I used high frequency stimulation of the pituitary gland causing the hormone secretions to reverse the growth process." Dr. Linstrom then adds, "first injections of the sulfa hydral compounds into the bone marrow." Dr. Coulter responds with, "that will stop the growing." It appears that in the film sulfa hydral is used to treat bone marrow and pituitary secretions thereby controlling an animal's size. The reality of an antibacterial is a long ways from controlling an animal's size through hormones.

Frankenstein's Daughter (1958): Digenerol

Dr. Carter Morton (Felix Locher) needs the synthetic drug Digenerol to complete his formula created for preserving cells and tissues. Morton shares his aspirations with Oliver Frank (Donald Murphy) a.k.a, Oliver Frankenstein, who unbeknownst to Morton is the grandson of *the* Dr. Frankenstein: "Think what it would be like. To be able to wipe out all destructive cells and organisms that plague man. No disease. No destructive tissue or growth. Man would be ageless." Oliver Frankenstein replies, "I'm afraid you're on the wrong track. Your formula may work on internal cells [called endothelial cells] but it causes violent disfigurement to sensitive exposed skin areas."

In one exchange, Morton says to Oliver, "We'll start the day by using acid pentyl to hold the formula together." Though it certainly sounds important, its meaning and

9. Drugs in Science Fiction Cinema

use are unclear. Compounds that are used to "hold formulas together" are collectively referred to as incipients and acid pentyl could serve this function. Acid pentyl, more commonly known as amyl acetate and a derivative of acetic acid is primarily used to solubilize organic molecules suggesting that the drug Morton is trying to develop is a small organic compound.

The developer of the drug warns, "We've been trying to perfect Digenerol for a number of years ... right now that drug is very dangerous ... it hasn't been perfected yet. In the wrong hands it could seriously disfigure anyone who might take it internally."

Unearthly Stranger (1963): Trimorphonite

At the "Royal Institute for Space Research" aliens are killing off the scientists. Traces of Trimorphonite a "powerful and totally immediate sedative" were found in the arteries of the dead scientists. Trimorphonite is "part of the morphine family," and used to subjugate and to behaviorally control people. A woman dosed with the drug is able to lift a hot casserole pot from a 275 degree oven with her bare hands. Though the drug would deaden the pain receptors, skin damage would still occur due to the excessive heat.

A Clockwork Orange (1971): Moloko Plus

Distributed in the Korova Milk Bar are several stimulants that are mixed with a substance called Milk Plus, available in three types, Vellocet, Drencrom, and Synthemesc. These stimulant drugs are used by Alex (Malcom McDowell) and his droog friends to go on psycho rampage benders. These drugs are similar to PCP and Angel Dust (amphetamine) in that they produce acute alertness and violent tendencies. As Alex describes, it "sharpens you up and make you ready for a bit of the ol' ultraviolence." Moloko Plus is a combination of all three: Vellocet (barbiturates or opium-related), Drencrom (adrenachrome), and Synthemesc (mescaline). Barbiturates, opium, and mescaline mixed together would make an interesting cocktail and certainly prepare you for some "ol' ultraviolence."

Why the drug needs to be taken with milk is interesting. Perhaps the many biomolecules and natural chemicals in milk help stabilize the drug and help it get absorbed through the digestive system and into the rest of the body. A secondary benefit could be a pseudo "slow-release" like effect in that the milk helps delay some of the absorption processes making the effects of the drug longer lasting since it slowly gets into the blood system. A result would be a longer lasting high. Of course, if one of the droogs were lactose intolerant then milk plus would be a problem.

But, "Science has cure!" screams a newspaper headline. To counter the trend of Moloko Plus an experimental drug is concocted, referred to as "Serum 114," it causes severe discomfort when violent tendencies strike. When a criminal is about to commit violence, the perpetrator is dosed with physiological countermeasures of pain, which take over and increase in intensity, stopping the criminal act before it happens.

Physiology

The Andromeda Strain (1971): Kalocin

Dr. Jeremy Stone (Arthur Hill), a Nobel Prize-winning bacteriologist is the developer of Kalocin, a super-drug intended to cure all diseases, known and unknown, including every known virus, bacterium, fungus, parasite, and cancer. Unfortunately, those who stop taking the drug quickly die from opportunistic superinfections that prey on the natural immune system compromised by Kalocin.

Kalocin came from the mind Michael Crichton, author of *The Andromeda Strain*. In pursuit of authenticity Crichton named the fictitious company that manufactured Kalocin, Jensen Pharmaceuticals, and made reference to the science journal, *Nature*, indicating the actual source of the drug. Of course there is no single drug that has the ability to cure all diseases, so Kalocin is alack and alas pure fiction.

Logan's Run (1976): Muscle

Muscle is a super-amphetamine that overstimulates neurotransmitter release in the brain and speeds everything up. The drug is briefly used on Logan (Michael York) while he is running to find Sanctuary. A cloth saturated in Muscle is placed over his mouth thereby disabling him so it can be considered an inhalant. Nevertheless, the drug wasn't used long enough for it to have any real effect on Logan. It would be an interesting contest to compare Muscle with Polydichloric Euthimal (see *Outland*), both amphetamines. As uppers, compare Muscle, the "super-amphetamine" with Polydichloric Euthimal, "the strongest thing you ever saw."

Also, during an early scene in the film a pleasure drug in a glass ball was thrown up against the ceiling causing it to break releasing a thick red cloud of dust apparently bringing about a mass orgy. This unnamed drug would enhance libidos as well as provide a sense of euphoria.

Saturn 3 (1980): Blue Dreamers

A Blue Dreamer is a synthetic drug that is given to scientists working at a hydroponics lab on an isolated moon of Saturn, "to keep them from going static," basically to alleviate boredom and other effects of long-term solitude. A blue dreamer is a small blue gel-paste tablet that is easily cut with a knife and taken orally with a liquid (appears to be tea). The drug enhances libido and the overall erotic experience and creates hallucinogenic perceptions during dream states.

Scanners (1980): Ephemerol

In the film, Ephemerol was originally given to pregnant women during the mid-1940s by Dr. Frane (Victor Knight), as a way to relieve morning sickness, but resulted in the birth of scanners, humans endowed with powerful psychic abilities.

9. Drugs in Science Fiction Cinema

Dr. Paul Ruth (Patrick McGoohan), a psychopharmacologist and the inventor of the drug gives his pregnant wife a prototype of Ephemerol and she has twin sons born as scanners.

When given to scanners Ephemerol acts as a suppressant. As Dr. Ruth explains, "It does nothing when given to ordinary human beings. When given to a scanner it prevents the flow of telepathy. It stops the voices." The drug interferes with cognitive senses and neural network.

The film flashes on the July 22, 1946, issue of *LIFE* magazine with a fictitious full-page ad for Ephemerol, perhaps a reference to real-life Thalidomide, which was a sleeping pill that also helped morning sickness in pregnant women primarily prescribed during the 1940s and 1950s. Unfortunately, multiple birth defects resulted, so the drug was pulled from the market.

The Empire Strikes Back (1980): Bacta

According to *Star Wars* lore, Bacta is a synthetic chemical consisting of a gelatinous fluid containing bacterial particles that promote wound healing and rapid skin tissue regeneration so it therefore has growth hormone-like properties. Bacta is used for everything from treating superficial injuries like minor cuts to whole body re-engineering as happens in *The Empire Strikes Back*. Since these injuries were superficial and not internal though severe, a different means of healing was needed such as a drug that focused on skin growth and integrity.

Luke Skywalker (Mark Hamill) is immersed in a tank of Bacta after suffering injuries from an attack by the ice creature, Wampa, and frostbite from the extreme cold temperatures of the ice planet, Hoth. The Bacta is dissolved in a serum resembling plasma, fluid that mimic the body's own, and allows all skin to be equally bathed in the drug, resulting in Skywalker's complete rejuvenation while also preventing scar tissue from forming.

Outland (1981): Polydichloric Euthimal

Polydichloric Euthimal is an illegal drug, promoted as a "tranquilizer ... company issue," to alleviate the boredom of working at a mining colony on a moon of Jupiter. As Dr. Lazarus (Frances Sternhagen) explains to Marshall O'Niel (Sean Connery), "It's an amphetamine. Strongest thing you ever saw. Makes you feel wonderful. You do 14 hours of work in 6 hours. It makes you work like a horse." Yet a blood sample has been taken from an apparent victim of suicide and reveals traces of Polydichloric Euthimal. The drug has side effects as Lazarus explains, "It also makes you psychotic. It takes a while ... 10, maybe 11 months. And then it fries your brain."

Amphetamines are a class of psychostimulant drugs that produce increased wakefulness while decreasing fatigue and appetite. Specifically, amphetamines increase neurotransmitter activity in the brain through the release of dopamine and norepinephrine hormones. This activity also brings about an elevation of overall mood, a heightened libido, and a sense of euphoria.

As with any drug, repeated high-dose intake can lead to a variety of mental states including delusions, psychosis, and paranoia. In *Outland*, the effects of Polydichloric Euthimal build up and affect some more than others, suggesting some fundamental psychotic tendencies must be present that are exacerbated by the drug.

Return of the Jedi (1983): Marca Herbs

This drug can be considered a natural product since it is derived from a plant and not apparently synthetically made.

When we first see Jabba the Hut he is smoking Marca Herbs with what appears to be a hookah. The water in the device is being aerated with bubbles and smoke encircles Jabba. The drug is an analog of marijuana and perhaps opium that in the *Star Wars* series the Huts seem to use as a cultural and religious practice. After smoking on his hookah the look on Jabba's face is one of relaxation and euphoria, complete with puffy eyes.

Dune (1984): Melange

Melange, also known as the "spice" is only found on the planet, Arrakis, also known as Dune. Melange is a natural drug derived from secretions of the giant sandworms that inhabit Dune. The amazing drug extends lifespans, grants greater vitality (increases adrenaline and testosterone), and makes interstellar travel possible. It even gives some people the ability to see the future by unlocking prescience. Unfortunately, there are always side effects, and beyond the run of the mill dangers of addiction, users risk the sclera of their eyes turning blue, and with high doses may undergo morphological changes that will make them fish-like as well as gain weight. Also, alcohol seems to enhance the effects of melange.

The Serpent & the Rainbow (1987): Zombinol

A zombified man who "displayed a negative pulse, no heartbeat, no respirations, no pupil dilation, no brain waves, no response to pain" was buried, only to be discovered alive years later. The head of the drug corporation researching Zombinol says, "Somebody brought him back from the grave and I want to know how they did it."

Back at the lab the effects of a sample of the zombie powder are tested on a baboon. This drug is inhaled so it has to be absorbed through the pulmonary (lung) system. Neuromuscular function is observed with measurement of the limbic, sensory, autonomic, and motor responses. Shortly after receiving the drug, the baboon's respiration, pulse, and blood pressure registers "zero." A scientist notes that while the physical body is paralyzed "the areas of the brain that controls sensation, thoughts and emotions are still active." Meaning, that on Zombinal ... "you could hear, see, maybe even feel, think, but you couldn't do anything about it." Even if you were buried alive. In the film the effects of the zombie powder wore off after 12 hours, or so, leaving the victim completely normal again.

As explained by the scientist, "the ingredients to the powder are terrifically varied. There is the poisonous sea toad, *Bufo marinus* [cane toad], the same animal Lucritia Borgia used. Made even more toxic by frightening it with a stinging sea worm. And the puffer fish that produces one of Nature's most powerful poisons, tetrodotoxin. Plus a whole pharmacy of herbs, minerals, charred and ground and mixed with a skill that is astonishing ... the process takes 3 days and nights." A complex procedure indeed and one that would be difficult to exactly reproduce each and every time with no variation.

Ethnobotanist, Dr. Wade Davis, is the author of the book, *The Serpent and the Rainbow*, a fictionalized account of his research on voodoo and zombies in the Caribbean. The actual chemical Davis discovered as the basis of zombification is tetrodotoxin, a real deadly toxin obtained from puffer fish. Tetrodotoxin is an inhibitor of nerve cell responses and acts on both the central and peripheral nervous systems that results in loss of respiration and vasomotor control. Just a small dose of 1–2 milligrams is enough to kill a human.

Robocop 2 (1990): Nuke

Nuke is a designer drug, synthetically made and controlled by Omni Consumer Products. They promise that those who take the narcotic, "...will know paradise every moment of their lives. There is a nuke for every mood. We will offer our customers every opportunity to control every aspect of their emotional lives."

Nuke for every mood includes varieties known as red ramrod, white noise, black thunder, and blue velvet. The most common form of Nuke is red ramrod, which is a red liquid enclosed in a single use, injectable vial. Each vial dose is about one milliliter of volume and is self-administered in the neck, via the jugular vein. In this way the drug will immediately circulate to the brain and the rest of the body thereby satisfying and intensifying the addictive craving.

"Thank you for your cooperation" a voiceover frequently announces on behalf of the mega powerful Omni Consumer Products. In the film Nuke is considered "the greatest health threat facing our nation," as even a single dose is enough to cause addiction and create a lifetime of dependence on the manufacturer.

Screamers (1995): Red

The surface of the planet, Sirius 6B, is heavily contaminated with radioactivity. To help neutralize the effects inhabitants occasionally take the drug, "Red," distributed in the form of a cigarette. When a large cloud of radioactive particles approaches a base the people at a crowded nightclub are instructed to light up and a voiceover announces, "radiation alert ... please light your Reds." Ironic to say "smoke your cure" which prompts the insightful reaction, "I can't believe you gotta put this shit in your lungs just to neutralize the shit in your lungs." When asked how one knows if the drug is working the response is a terse, "you don't die." Frequently lighting cigarettes suggests the drug has a temporary effect and not a long lasting one.

Inhalation therapy is a common method to administer drugs, especially to patients with lung problems. For inhalation therapy to be effective in neutralizing radiation the drug would most likely be a chelator, a binding agent that grabs each radioactive atom in a cage-like structure, thereby making it easier for the body to eliminate and not be absorbed.

Dark Angel a.k.a., *I Come in Peace* (1997): Blarcy

Aliens come to earth in search of Blarcy, a recreational drug made of human endorphins. To obtain Blarcy, an alien injects humans with a high dose of heroin which causes a massive release of endorphins, which the alien extracts and subsequently uses as his drug. It was stated that "one ounce would be enough for one thousand doses." An ounce is 28.35 grams and dividing it by 1,000 would make 28.35 milligrams per dose. A large dose for anyone that would more than achieve the "high" sought after.

As a coroner states, "Heroin stimulates the pituitary to make endorphins which are hormones that create an incredible sense of well being. Nature's ecstasy."

Natural drug compounds are "natural" to the endogenous species but can be highly reactive in another species. Many of the products harvested from plants are natural to the plant species but can have quite dramatic effects in different species, such as man, heroin and opium being good examples.

The Matrix (1999): Red Pills or Blue Pills

Choose Red or Blue.

The Blue Pill provides the blissful ignorance of illusion (the fabricated Matrix reality) whereas the Red Pill reveals the painful truth of reality (the real world). These drugs must be anti-psychotic and LSD-related to cause such strong mental images in those who take it.

Upon taking the Blue Pill the person will lose consciousness and remain in the fabricated world of the Matrix, perhaps never wondering how deep the rabbit hole goes. The person taking the Red Pill will become unplugged or freed from the Matrix and live in the harsh real world. Each person gets only one chance to choose.

Minority Report (2002): Neuroin

The drug Neuroin appears to be a combination of Heroin and a neuro compound. Neuroin is taken via an inhaler so it can be considered a gas form of Heroin and

9. Drugs in Science Fiction Cinema

therefore huffable. The drug gets into the lungs and then directly into the bloodstream via the lung tissues. Use of the drug induces an extremely relaxed state of euphoria accompanied by a dulling of the senses. The drug can be considered a psychotropic in that it alters mood as well as cognition and perception, and maybe even leads to enlightenment. The offspring of Neuroin addicted users have clairvoyant abilities and are used to forecast extreme crimes such as murder, to enable early intervention and prevent the crime from happening in the first place.

In addition to the above 20 drugs there are a few more fictional drugs that may be of interest. The drug "Hypnocil" is used in *A Nightmare on Elm Street* #3 to suppress dreams, though an overdose results in an irreversible coma. In the original TV series, *Star Trek*, in the episode, "Mudd's Women," a "Venus Drug" is used to enhance the appearance of women and to make men more aggressive. *Star Trek* also has hyronalin which counteracts radiation poisoning or cordrazine. There is "Soma" from Aldous Huxley's novel, *Brave New World*, a drug that keeps everyone sweet and blissful. Also, "Substance-D" from Philip K. Dick's, *A Scanner Darkly*, a drug that essentially twists your mind in a way that only Philip K. Dick can. Then there are "Mindjacks" from *Strange Days* and "Tek" from *Tekwar* and Zombrex—ambuzol vasiplatin from the popular *Dead Rising* comic book.

In these classics of SF cinema, drugs come in all shapes and sizes, flavors and colors, and methods of administration, including those that are ingestible, inhalable, or injectable. And their powers are extraordinary.

Summary

The films discussed here span a time frame of about 70 years and during this time much has changed about our understanding of what drugs are, what they can do, and ways to use them to improve life and health. "Better living through chemistry" is the unstated motto of these films. Though the extraordinary drugs used in SF cinema have interesting and unique names they essentially represent real drugs that act in similar ways. Basically, they are ordinary drugs with unusual names. Even those "one-of-a-kind" drugs uniquely made by our cinemascientist for a specific application work in relatively ordinary ways. And of course those drugs not of this earth in their own unique ways too. These SF drugs, both natural products and synthetic, are mostly hallucinogens, amphetamines, opioids, or hormone enhancers. All in all, both real and imagined drugs in the SF film world are of interest and help move our favorite plots along. And drugs help make our monsters and their creators scary.

10

Hormones, the Scariest of Them All!

My first exposure to the real world of hormones was while a sophomore in college where I was allowed to take an upper division endocrinology course. Much of it at the time was brute memorization since I did not have sufficient background to appreciate the nuances of hormones, but ongoing exposure led me to recognize the power and beauty of hormones.

At the core of the actions and behaviors of our favorite SF film monsters are hormones, which are the most powerful substances in our bodies. With such a rich and diverse variety of hormones I am surprised by how few of them have been exploited, directly or indirectly, in SF films. Most of the hormonal experiments are usually limited to either growth hormone, resulting in gigantism, or androgens, like testosterone, granting super-strength and virility.

Hormones (from the Greek, ὁρμή, meaning "impetus" or "set in motion") are a class of regulatory biochemicals produced by endocrine glands in our bodies and transported by the circulatory system to a distant target. The field of endocrinology is the study of hormones. Hormones regulate a variety of physiological and behavioral activities, including digestion, metabolism, respiration, tissue functions, sensory perception, sleep, excretion, lactation, stress, growth and development, movement, reproduction and mood. Generally, only a small amount of hormone is required to alter cell metabolism. Hormones are also a silent driver of behavior and personality.

Some examples of endocrine glands are the adrenals, thyroid, ovaries, pituitary, testes, thymus, and pancreas. Other examples are the hypothalamus, and pineal gland. Hormones are responsible for determining an animal's sex, size, shape, organ structure, and ability to respond to the environment. As a strict definition a hormone is any molecule made by one cell that is transported via the circulatory system to a different cell. And the endocrine system is so complex that the same hormone behaves radically differently from one person to the next. In animals, the brain is often a target organ for many of these hormones, and the brain, in turn, regulates the secretion of these hormones. And the rate of these hormonal changes and influences can also widely range from immediate to long term depending upon the specific hormone. Complex indeed.

Once a hormone is made (or synthesized) it gets transported through the cir-

10. Hormones, the Scariest of Them All!

Cover to *Scary Monsters* #97. The back cover plugs the article "Hormones, the Scariest of Them All!" ©2016 Dennis Druktenis Publishing & Mail Order, Inc.

culatory system to home in on its target, called a hormone receptor. If the hormone, such as a protein hormone like insulin, is water soluble it is readily transported through the blood stream. If the hormone is not water soluble like most steroid and thyroid hormones, then they must piggyback other biomolecules for transport through the circulatory system. Also, some hormones are completely active when released into the blood stream (like insulin) whereas others must be activated through a series of coordinated steps.

Hormone History

The early history of hormones had a shaky start. Initial efforts were targeted at rejuvenation and attempts to restore physical, mental, and sexual vigor. In 1889 a 72-year old physiologist, Charles Edouard Brown Sequard (1817–1894), reported an improvement of his frail heath after injecting himself with dog and guinea pig testicular extracts. Shortly afterward almost every conceivable tissue became the "cause celebre" for essentially every conceivable disorder. The focus was soon on the "mysterious secretions" of ductless glands, especially the sex glands, and serious experimental research was conducted. One popular medical publication from the 1920s stated, "that the abnormal functioning of these ductless glands may change a saint into a satyr; a beauty into a hag; a giant into a pitiful travesty of a human being; a hero into a coward; and an optimist into a misanthrope." It should be noted that all these above comments pertain to the down side of hormones. It could also be said that hormones can change a satyr into a saint, a hag into a beauty, and a coward into a hero.

During the heady endocrine research days of the 1920s, the expectations were that hormones would provide a revolutionary way to improve human life. A major part of this work at the time was the use of "monkey glands" or more specifically ape testicles. One prominent surgeon, Dr. Serge Voronoff (1866–1951), implanted ape testicles into aging men. Voronoff also claimed that by surgically grafting the glands of animals into other animals, including humans, he could correct some birth defects and help prolong life. Even though extreme, these innovative studies helped to establish the difference between male and female hormones and how they influenced male and female development.

How Hormones Work

Hormones work in tiny amounts, so small changes can effect major changes in the body. Even a slight excess or a slight deficiency of a hormone can lead to disease states. In mammals, including humans, most protein hormones bind to a specific membrane-embedded receptor on its target tissue which in turn elicits specific cellular events to occur. A hormone binding to a specific receptor is very much like a certain key that fits a specific lock. A hormone can only work if its specific receptor is present on the target tissue. And the converse is also true in that the target tissue can have many receptors but without the specific hormone to activate it, it remains unresponsive. Also, a cell may have several different receptors that can either recognize the same hormone or several different receptors for completely different hormones. The specific hormone receptors can be found on many different types of tissues such as with insulin where the insulin receptor is on a variety of cell types that triggers a diverse range of responses. These hormones can either all activate the same biochemical pathways or activate different pathways for different physiological responses. Hormones are the ultimate regulators.

Once a hormone binds to its specific receptor a series of protein triggers are activated by what are collectively called signal transduction mechanisms (via a cascade of biochemical events) that leads to the activation of a series of genes that control cell and tissue physiology. This could also include the production of additional hormone receptors that in turn could bind even more hormones thereby making an even greater biological response. Overall, hormones tend to increase, or up-regulate, protein synthesis in cells.

For non-protein hormones such as steroid or thyroid hormones their receptors are typically located within their target cells (a cell area called the cytoplasm) so they must cross the cell membrane to enter. This combined hormone-receptor complex then binds to certain DNA genes to either activate or suppress them.

Hormonal Signaling

Signaling of a hormone involves the biosynthesis of the particular hormone, its storage and secretion, the transport of the hormone to its target cell(s), the recognition of the hormone by its specific receptor, the relaying and amplification of the hormonal signal (usually the binding of the hormone to its receptor) which in turn leads to a defined cellular response. Lastly is the degradation of the hormone when it is done doing its job. An important consideration here is the concentration of the hormone-receptor complex which is determined by three factors: physiology, effects, and structure. The concentration affects the level of signaling and therefore the level and intensity of the response. Furthermore, there are also natural feedback mechanisms in place that help maintain a homeostasis of hormonal balance so when an overproduction occurs it can be naturally dissipated or neutralized.

Physiological Effects

Hormones are defined functionally, not structurally, so they may have diverse chemical structures even though they work on the same target. In mammals hormones have the following effects: stimulation or inhibition of overall growth, wake-sleep cycles both circadian (24 hr. cycles) and hemi-circadian (12 hr. cycles) rhythms, mood swings, control of programmed cell death, control of the immune system, regulation of metabolism, hunger cravings, puberty, mating, menopause, and control of the reproductive cycle.

Hormone secretion can be stimulated or inhibited by other hormones, by the concentration of nutrients in blood, by neural activity (including mental activity), and by environmental changes (such as with temperature or light). In addition there are both natural and synthetic chemical compounds that have hormone-like effects. At times these compounds change the body homeostasis as effectively as natural hormones.

In our bodies the effects caused by hormones are concentration dependent meaning the more the hormone then the more the effect. Also involved in behavior issues are the release patterns of the hormones, the location of the hormone receptor, as well as their efficiency in bringing about changes; some occur in seconds whereas others may take years to see effects. Furthermore, the environment also highly influences hormone expression and function. A feedback mechanism works in that behavior influences hormone expression and hormone expression influences behavior.

Physiology

Chemical Classes of Hormones

In animals, including humans, hormones are of three major categories: peptide, lipid, or monoamine. Peptide hormones are made of chains of amino acids, examples being thyroid releasing hormone, endorphins, and vasopressin. Longer peptides, upwards of hundreds of amino acids are considered proteins, such as insulin and growth hormone. The main class of lipid hormones are steroids such as estrogen, testosterone and cortisol. Another example are the prostaglandins that are involved in smooth muscle contraction and relaxation. The monoamines are all derived from amino acids like phenylalanine, tyrosine, and tryptophan, and function as neurotransmitters.

Hormones and Diseases

Diseases and disorders of hormones can be readily diagnosed both in the clinic as well as the laboratory. There are many lab tests available that can accurately measure levels of hormones in body fluids like blood, urine, and saliva to monitor hormone health. In some cases, hormone replacement therapy is used to compensate for deficiencies and some medications are available to inhibit overproduction of other hormones. For example, those deficient in insulin production, like diabetics, can receive insulin replacement therapy and those with an overactive thyroid can be treated with a synthetic drug (e.g., propranolol) that counteracts the effects of excess thyroid hormone.

Many hormones and their synthetic analogs are used for medication with the most common being estrogens, progestogens (birth control), thyroxine (for hypothyroids), and steroids (mostly for autoimmune diseases). The hormone insulin is used for diabetics. Typically a pharmaceutical dose of a given hormone is usually much larger than what naturally occurs in the body so the overall response may also be different and beneficial. However, it should be noted that hormone medications do have some potentially harmful side effects.

Pineal Gland

The pineal gland is a small endocrine gland located in the brain and produces melatonin, a hormone that affects sleep patterns in circadian rhythms. It shape resembles a pine cone (where its name came from) and is about the size of a grain of rice and located in the epithalamus near the center of the brain. (Just so you know, the alligator lacks a pineal gland; see *The Alligator People*.)

The philosopher, Rene Descartes (1596–1650), believed the pineal gland to be the "principle seat of the soul." Its location deep within the brain suggested to philosophers

10. Hormones, the Scariest of Them All!

that it must have some significance. This is why it has been regarded as somewhat of a "mystery gland" with mystical and occult overtones about its perceived function. Another popular term for the pineal gland is the "third eye" since it has been compared to the photoreceptive parietal eye present in some animal species. There is some support for this since pineal cells possess a common evolutionary ancestor with retinal (eye) cells. The pineal gland was originally believed to be a vestigial remnant of a larger organ so there has been a mystique about this gland for some time. Overall, the hormonal substance from pineal glands, melatonin, has not been proven to be especially useful.

Melatonin is a derivative of the amino acid, tryptophan and its production by the pineal gland is stimulated by darkness and inhibited by light, all a part of the body's natural circadian rhythms. When puberty arrives the production of melatonin is reduced so it seems that pineal gland secretions inhibit the development of reproductive glands (see *The Leech Woman*). In addition to its role in sexual development the pineal gland is also involved in hibernation in animals, their metabolism and seasonal breeding. As humans approach adulthood the pineal gland undergoes calcification and continues this process during aging. Some think the pineal gland calcification contributes to Alzheimer's disease.

Pituitary Gland

In humans, the pituitary gland or hypophysis is a hormone gland about the size of a pea, though located at the bottom of the hypothalamus in the lower center of the brain, it is not a part of the brain. The pituitary gland consists of two components, the anterior pituitary and the posterior pituitary, and in total secretes 9 different hormones. The hypothalamus releases various growth factors to the pituitary gland which in turn stimulates the release of the pituitary hormones. As such, the pituitary is known as the "master endocrine gland." Some of the hormones the pituitary gland secretes are the somatotrophins (also known as HGH or human growth hormone), thyroid-stimulating hormone (TSH), endorphins (see: *Dark Angel*), prolactin, melanocyte-stimulating hormone (influences skin pigmentation), vasopressin, and oxytocin (increases labor contractions in women).

Phytohormones

Plant hormones are known as phytohormones which are chemicals or signal molecules that regulate plant growth. Plant hormones determine the formation of flowers, stems, leaves (and the shedding of leaves), and the development and ripening of fruit. Plant hormones affect seed growth, time of flowering, the sex of flowers, senescence of leaves and fruits. Plant hormones are simple chemicals and not the complex biomolecules seen in mammals. It should be noted that not all plants respond to hormones.

Classes of Plant Hormones

There are five classes of plant hormones that can vary from plant to plant and their chemical structures are significantly different suggesting specific and defined actions. Each class has positive as well as inhibitory functions that often work in tandem. The five major classes are: abscisic acid (an important plant growth regulator produced in leaves, especially when plants are under stress), auxins (influence cell enlargement and bud formation), cytokinins (a group of chemicals that influence cell division and shoot formation), ethylene (a gas produced as a metabolic breakdown product that influences leaf expansion, stem diameter and height), and gibberellins (a large range of chemicals important in seed germination, promoting flowering, and growth of new cells). Though there are other known plant hormones they are highly specialized for specific plants and regulate defense mechanisms, pollen compatibility, shoot branching, and nitrogen fixation.

There are also synthetic plant hormones that are used for plant propagation from cuttings, grafting, and various tissue culture procedures.

Hormones Made Me Do It!

In many SF films the drug or "treatment" by the scientist ultimately works via hormones. The drug acts as a messenger that sends or stimulates a particular biological signal for other things to happen. For example, all the SF drugs or radiation exposures that selectively stimulates the production of growth hormone (plant, insect, or animal/human) results in a proportionately uniform response.

One major flaw in this is how to *selectively* direct the hormone production to either certain species and/or certain parts of a particular organism, irrespective of its size. Overall uniform growth, such as in gigantism, may be a concept that is easy to understand, but selective growth either in a population (such as seen in *Them!*) or in an individual (Colonel Glenn Manning in *The Amazing Colossal Man*) is not so easy. Complicated biochemical and genetic programming would be necessary to achieve the results portrayed in these films.

Many of the behaviors of our favorite screen monsters can be attributed to raging hormones. And since hormones work in minuscule amounts it doesn't take much to influence behavior. Mood swings are often due to hormones gone wild.

And if someone or something large has a large amount of hormones coursing through their bodies, then inevitably much mayhem will result. No wonder many of our favorite cinemonsters are angry ... their hormones are out of balance!

10. Hormones, the Scariest of Them All!

Glands and Early Cinema

During the early part of the 20th century when the film industry was just getting its legs popular science was fixated on glands and gland extracts and how they could "restore youth" and "bring vigor back to life" (the proverbial "Fountain of Youth" is nothing more than a bath of hormones). This reflected the new science of endocrinology at the time when many of the hormones were beginning to be discovered and this pop science was brought to films as plot devices.

During the 1920s through the 1930s the use of gland extracts was popular fodder for film plots as seen in *Murders in the Rue Morgue, The Monster Maker,* and *The Ape Man.*

During the early days of SF cinema it was all the rage to have mad scientists conducting glandular and hormonal experiments and to talk about "glands" and what they can do to greatly improve health and well being.

At that time the "restoration of youth" was simply a "promise" for older men to regain their youth by being able to have sex. In other words, the use of testosterone (simply thought of as male hormones at the time) could make old men virile again like they were during early manhood.

After it was understood that hormones are produced from glands the cinema concepts changed to incorporate that understanding. Though the plot drivers were the same, namely the use of biological material to affect certain changes, the means to do so have changed. As the public's understanding of science became more sophisticated after World War II the film plots matched this sophistication and began to use more complicated biology and hormones were a key art of this.

At the end of World War II when biomedical research had advanced, the word "hormone" began to appear in film scripts. In the 1942 film, *The Corpse Vanishes,* the phrase "glands and hormones" is said, and that may be the first use of the word "hormone," in SF film. This demonstrates that early 1942 script writers, such as Harvey Gates, were up-to-date with science.

The key moment in *The Corpse Vanishes* when the female reporter says "glands and hormones." This may be the first time the word hormone was used in SF cinema. (Left to right, Kenneth Harlan, Luana Walters, Tristram Coffin.)

Physiology

In SF films most of the hormonal effects, either directly or indirectly implied, are somewhat limited to either growth hormone (gigantism) or androgens like testosterone (superstrength and virility). The myriad of other hormones are mostly ignored in SF cinema. Even a condition as common as diabetes is limited in scope in SF films. One possible interpretation is no real monster can be created through the use of most hormones. Imagine "The Diabetes Monster," or "The Thyroid Terror," or possibly "The Hypothalamus Horror" and it may be difficult to get producers interested.

The Films

Dr. Jekyll and Mr. Hyde (1932)

Various hormones.

There are many filmed versions of the Robert Louis Stevenson classic and one of the best is the 1932 version starring Frederic March, who won an Academy Award for his performance. The potion that caused the changes from Jekyll to Hyde induced a massive hormonal imbalance resulting in many physical and emotional changes. Many hormones were affected including those that regulate skin tone, teeth, hair growth, as well as behavioral changes. His libido became dominant (more testosterone) and he no longer felt compelled to suppress any urges. At first the hormonal changes were temporary and Jekyll reverted back to Hyde after consuming a counter-potion, which would be a hormone suppressor. Each successive transformation into Hyde (there were 5 in total) brought more severe physical and emotional distortions. After acclimating to the process, his brain eventually took over and spontaneously released a storm of hormones bypassing the need for an external trigger to induce the transformation.

Island of Lost Souls (1933)

Growth hormones; estrogen.

"What is the law?" intones the Sayer of the Law (Bela Lugosi). Well, the law is actually hormones. Doctor Moreau (Charles Laughton) uses a combination of surgery and hormonal manipulation to transform animals into manimals. Since hormones work in a transient manner they must be re-administered to maintain their effects (think of a diabetic who must take the hormone insulin every day) which is why Lota (Kathleen Burke), a panther woman, Moreau's "most nearly perfect creation," reverts to her natural state with, "stubborn beast flesh ... creeping back," when the hormone treatments stop.

The Ape (1940)

Nerve growth factor.

Dr. Bernard Adrian (Boris Karloff) uses spinal fluids, to help a paralyzed girl,

10. Hormones, the Scariest of Them All!

Frances Clifford (Maris Wrixon), walk again. Dr. Adrian initially uses spinal fluid from an ape, which at first works since human and ape proteins are about 98.5 percent similar, but the effects turn out to be only temporary, so he turns his search for spinal fluid to humans, who don't usually survive the extraction.

In the film the purpose of the spinal fluid was to cause growth of new nerve tissues that would repair damaged muscle coordination and allow paralyzed people to walk again. The gorilla was provided via an accident at a local circus and the animal's spinal serum was successful enough it gave Adrian the impetus to seek out human spinal fluid. In reality, what Adrian wanted was a protein hormone called nerve growth factor that would help stimulate nerve growth. In the search for nerve growth factor hormone this protein can be obtained from a number of cell types and tissues and not specifically from spinal fluid.

The Mad Monster (1942)

Growth hormone, testosterone.

Mad scientist Dr. Lorenzo Cameron (George Zucco) has been thoroughly discredited by his academic colleagues, who comment, "Your crazy experiments are a disgrace to science." In addition, a professor adds, "What good can come from tampering with the normal laws of nature? ... Mingling the blood of man and beast is down right sacrilege."

His goal is to build "an army of wolf men."

In Cameron's lab we see a "wolf," looking suspiciously like a coyote, in a cage with a gauze around its left front leg, simulating a blood transfusion procedure. The blood is drawn into a glass container. Cameron mixes a small amount of the blood with a liquid buffer and injects it intravenously into his assistant, Petro (Glenn Strange), who rapidly transforms into a werewolf. As such, no full moon is required for the transformation. With such a rapid change then the hormone effects were on tissues and muscle movements and not on a more fundamental change at the genetic (DNA) level.

After concocting an antidote, Cameron excitedly explains, "I've discovered a certain extremely volatile element in the blood that are more than electrical particles of energy. A source of all physical growth and mentality. By exciting the various glands and brain cells I've learned how to extract and concentrate these elements from the blood of various animals. I can control evolution. I've discovered the source of life." Hormones, of course, are the "volatile element" that can cause dramatic changes at both the physical and mental levels. Unfortunately, wolf hormones will only work on fellow wolves and not on humans.

Dr. Renault's Secret (1942)

Growth hormone, testosterone.

Dr. Renault transforms an ape into a human using a combination of surgery and hormonal therapy. If the transformation was done at the genetic level the changes would be permanent, but since they were enacted at the cellular level, via hormones, then the changes would be temporary, and after the hormones wore off, so would the effects.

Renault says to Noel, "For years I've worked to change your appearance, to make you talk." Noel responds, "Want to make me like man." Renault describes it as, "an experiment in transmutation," saying, "I have undertaken glandular injections and brain surgery ... the nerve graft operations were a success. The electroencephalogram indicates that the brain traces are becoming more human each day.... The speech in the left hemisphere is fully developed.... I've proved my theory." Ultimately Noel "reverts back to an animal state," killing Renault suggesting that the animal state is one of wanton killing. In the wild, apes do not kill.

The Corpse Vanishes (1942)

Estrogen.

Dr. Lorenz (Bela Lugosi) uses estrogen extracts taken from young women (brides-to-be) to bring about regenerated youth in his elderly wife. To obtain these extracts Lorenz inserts a syringe behind the right ear of each victim and removes about 5cc fluid. Lorenz then adds the fluid to another flask and adds more liquid. From this he removes about 5cc and injects it into his aged wife behind her right ear. Shortly afterwards her youthful beauty is restored. The effects are temporary, lasting only a few hours, creating a need for a continuous supply of virginal estrogen.

Dr. Foster, a colleague, says, "trying to find a cure for his (Lorenz) wife ... although she has the appearance of a young woman her heart and arteries indicate that she is at least 70 to 80 years old." Foster further comments about Lorenz's work, "sustain his (Lorenz) wife in a youthful state." A female reporter comments, "scientists are finding out every day that glands and hormones have a lot to do with life and health." Foster adds, "Glands in our bodies help determine the condition of our teeth, the texture of our hair." Though odd examples of how hormones can work and not what most people would first think of they nevertheless are accurate.

The Ape Man (1943)

Growth hormone; nerve growth factor.

Dr. James Brewster (Bela Lugosi) creates an ape serum that transforms him into an ape man. The serum was likely hormone-based and came from either body fluids, including plasma, spinal fluids, or seminal fluid extracts, all of which influence hormone production and activity. The morphological ape changes Brewster undergoes are superficial and include hunched shoulders, which make his arms appear longer, hands bent at the knuckles, and excessive hair growth on his face, chest, and hands. He looks like a chimpanzee, but he acts and talks like a human.

10. Hormones, the Scariest of Them All!

A hormone-based ape serum would work in a transient manner, so the effects would not be long-term, but Dr. Brewster's transformation appears to be permanent. Perhaps he became impatient and just needed to wait longer for the effects of the ape fluid to wear off. For permanent changes to occur the DNA of the organism must be changed, something no spinal fluid can do.

During one exchange Brewster tells a colleague, "I must have human spinal fluid injected into me. It's my only chance. It's the only way to counteract the ape fluid injections." (It was never stated whether these counteractions would be permanent or temporary.) The injections are given in Brewster's lower back spine. The injection could actually be given anywhere and not necessarily the spine itself. If the human spinal fluid only works in a temporary manner then why isn't the originally used ape fluid also temporary? To keep a supply on hand Brewster has a medium-sized jar filled with about 300ml of "spinal fluid" and says he has enough for six injections (the sterility of that jar is very suspect) so therefore each injection would be 50ml, quite a large volume, especially if it has to be injected into the spine (ouch!). After injecting himself with the spinal fluids Brewster's ape-man slouch and monkey mannerisms were eliminated though his hairiness remained (why not shave?). One has to ask if this process of reverting back to human was a physical or a mental process?

Captive Wild Woman (1943)

Famous endocrinologist, Dr. Sigmund Walters (John Carradine), is conducting experiments with gland transplants. In his office at the Crestview Sanatorium, Beth Colman (Evelyn Ankers) reads from the fictitious medical journal *Medicine and Surgery* with an article by Dr. Walters, demonstrating to all that he is indeed a world expert on all things glandular.

Later, Walters declares, "It's been proven beyond doubt that glands can transform physical matter into any size, shape or appearance. We have here in this Sanatorium a rare case of a follicular cyst which induces the secretion of unusual amounts of the sex hormone." Then to his nurse, Walters says, "What's to prevent us from transfusing these glandular extractions from a human being into a higher type of animal?" The nurse's answer of "The authorities!" goes unheeded. A follicular cyst is not that rare in women and can form when ovulation does not occur and an ovary follicle does not rupture upon ovulation but continues to grow thereby becoming a cyst. Some cyst follicles can grow to about 6 cm (2–3 inches) in diameter and their eventual rupture can cause a sharp, severe pain on the side of the ovary where the cyst formed. Usually after a few months the cysts themselves disappear with no symptoms.

Walters obtains a female gorilla, Cheela, from a circus and transfuses human female glands and blood into the gorilla transforming the primate into "Paula Dupree" (played by Aquanetta). In the film we see the gorilla's left hand transform into a female human hand, as the hair disappears, the fingers become less round, and the flesh tone lightens. It is assumed that female glands with estrogen are what caused the transformation into a human female. It should be noted that both gorillas and humans have the same estrogen hormone so it is unclear how such a common hormone could have transformed an ape into a human.

To further complete the transformation of the gorilla into the woman Paula Dupree, Walters needs much more than mere female glands, "a cerebrum is essential to the success of this experiment." Needless to say, in donating her cerebrum the human donor, Dorothy Coleman (Martha Vickers), would die. Walters comments to his nurse, "Why should a single life be considered so important? ... She (Dorothy) will die in the advancement of science." The transplant, a procedure that would ordinarily be done by a team of surgeons, was performed solely by Walters in a mere two hours and 40 minutes. After the successful operation the nurse describes, "a human form with animal instincts," indicating even though Paula may look human the instincts are still those of a gorilla. Also, the gorilla's brain would have to be removed to accommodate the human cerebrum. The supposition is that a human brain would secrete human hormones thereby transforming an ape body into a human body. Stating the obvious, transplanting a human cerebrum into a gorilla would have no effect.

Walters uses imprinting and hypnosis to control the animal instincts of Paula Dupree. To test her humanity Walters permits her to work at a circus where she assists with the lions and tigers by using her "gorilla senses" to control the animals. Meanwhile Paula proves all too human and falls in love with lion trainer Fred Mason (Milburn Stone). When she witnesses Fred hug and kiss his fiancée, Paula undergoes a "hormonal fit" that results in her transforming back into Cheela. Walters comments, "One thing I hadn't counted on, a terrific emotion would destroy the new tissues in your gland growths ... now I have to do it all over again ... another brain transplant."

The Monster Maker (1944)

Growth hormone.

Dr. Igor Markoff (J. Carrol Naish) is a "specialist in glandular disorders," who has perfected a serum called X54, which cures acromegaly. To make his X54 formula, he starts with a "concentrate of the pituitary." Boil dry, turn off the burners, add some elixir, place in four cc ampules, and refrigerate them. All in all a common procedure.

Regarding his work, Markoff says, "I have succeeded with X53 in arresting the disease acromegaly, but it will do no more than hold it in check. No change, neither progressive nor retrogressive. I have every reason to believe this new formula will be a complete cure of the disease. Think what it would mean to have the power to control a dreaded disease like acromegaly."

Serum X54 is a necessary antidote as Markoff also developed an acromegaly-inducing formula, containing human growth hormone, which he keeps in a brown bottle, labeled, "A-5-B2." When concert pianist and protective father Anthony Lawrence (Ralph Morgan) objects to Markoff's pursuit of his daughter Patricia

10. Hormones, the Scariest of Them All!

(Wanda McKay), Markoff injects him with 2cc of "A-5-B2" by syringe, which results in rapid onset acromegaly. Typically, it takes years for acromegaly to manifest itself, but Markoff now has a case to test his new serum. Lawrence accuses the doctor, "You set yourself up as a Frankenstein and created a monster. I am that monster!" Interestingly, the curse is "a long and slow treatment," whereas getting the disease by Markoff is very rapid.

Acromegaly, a glandular ailment, is a slow progression syndrome that occurs when the anterior pituitary gland produces excess growth hormone. Normally at puberty the production of growth hormone ceases but in those with acromegaly the hormone continues to be produced and most commonly affects adults in middle age. As a result of this condition flat bones such as the jaw, cheekbones, forehead, hands, and hip continue to grow throughout life resulting in a misshapen and sometimes severely disfigured body. Serious complications including premature death can occur if not properly treated. Since acromegaly is a slow progressing disease it is difficult to diagnose during its early stages and is often missed for many years. The actor, Rondo Hatton (*The Creeper* and *House of Horrors*) was afflicted with this in real life which is why he looked like he did. The Rondo Awards are named in his honor.

Mesa of Lost Women (1953)

Growth hormone; estrogen.

Dr. Aranya (Jackie Coogan) is an endocrine entomologist who uses the power of hormones to alter insects and humans. The narrative voiceover describes Dr. Aranya's experiments as creating, "strange women who do not die." While analyzing Aranya's work a visiting scientist expresses interest in learning more about his research. The entomologist shares his secrets, "I have isolated the growth hormone of the anterior pituitary, the specific substance which controls the growth pattern of human beings.... What would the effect be if this hormone or a complete human pituitary was transferred to the body of another creature?" He describes moderate success among the lesser animals and complete failure among birds, but "while experimenting with hexapods I came upon the Seracedia family. Tarantulas. The tarantulas began to yield amazing results. They began to grow as large as human beings ... and then I reversed the process by transplanting the control substance back into the human body. Observe this girl. I call her Tarantella. She has beauty and intelligence and still possesses the capacity and instincts of the giant spider ... she has the indestructibility of the insect! And if her body became damaged and lose an arm or leg, she would grow a new one."

The family name for tarantulas is Theraphosidae and has about 900 different species, none of them named Seracedia. Also, since spiders have eight legs and not six they should be called "octopods" and not hexapods. Lastly, spiders are *not* insects, they are arachnids.

Regarding Aranya's lab, he has copious glassware indicating working on small organic molecules. An anatomical wall chart and cases of identified insects and arachnids are a nice touch. Lastly, he has one cheesy microscope that is inefficient; it only has one (!) lens (a len!?).

The Amazing Colossal Man (1957)

Growth hormone.

Due to radiation poisoning Glenn Manning's (Glenn Langan) hormonal balance was thrown completely out of whack and his adult body began to grow again, eventually reaching the height of 60 feet. For such massive and sustained growth to occur a constant supply of growth hormone, not to mention all the nutrients necessary for such growth, would have to be produced. This means that Manning's pituitary gland was in hyperdrive with a constant stream of growth hormone that uniformly caused his body to grow to correct gigantic proportions.

Within the film the leading scientist, Dr. Linsrom, reasoned that if they could at least stop the production of growth hormone then they could also stop the growth. In trying to explain Manning's condition to his fiance the doctor says, "the body is like a factory, continually producing new cells to replace the older cells, damaged cells, or destroyed cells. Now this happens in all the different parts of the body. Bone cells grow new bone cells, skin cells grown new skin cells and so forth throughout the body. New cells replacing the damaged ones ... it is this delicately balanced process of new cells replacing dying cells or damaged cells that is causing the growth problem with Glen.... For some unknown reason new cells are growing at an accelerated or speeded up rate while at the same time the old cells are refusing to die. This is what makes Glen grow. That's what made the new skin." In the film, Linstrom thought that sulfa hydral drugs could stop the growth but not reduce it, which in reality could do neither.

The Beginning of The End (1957)

Plant hormones.

This film is a rare example in the SF genre in which plant hormones play an interesting role. At the greenhouse of the U.S. Department of Agriculture, Illinois Experimental Station, Dr. Ed Wainwright (Peter Graves) shows his enlarged tomatoes and strawberries to reporter Audrey Aimes (Peggy Castle), explaining that radiation induced acceleration of the plant growth hormones. With such large strawberries and tomatoes then the plant hormones would have to be constantly produced and be transported into the fruiting bodies to spur growth. Not mentioned is the large volume of water in these fruits which by their very nature would weigh many pounds and therefore cause the branches to bend and possibly break due to the excessive weight.

In explaining the large fruit Wainwright says, "This we hope is the future of the

10. Hormones, the Scariest of Them All!

American farmer and for that matter all farmers everywhere." The reporter responds "Can you eat them?" Wainwright says, "No. Not yet. But we hope to develop one day a hybrid that can be eaten." Then the reporter asks, "How do they get so big?" Wainwright says, "Radiation causes photosynthesis, that is the growing process, that continues night and day. The radioisotopes act as a sort of artificial sun, a sun that never sets." The idea here is radioactive isotopes cause plant hormone production. Regarding the soil he uses Wainwright says, "That's plant food and essential minerals. It keeps the plants from burning themselves up. They have to be fed constantly. Actually, the fruit would grow much larger if we didn't limit the stimulation." Alack and alas, more fiction than science.

She-Devil (1957)

Pineal gland, estrogens.

At the core of this film is the pineal gland and the physical and emotional changes brought about by a "serum" derived from a fruit fly (*Drosophila melanogaster*). Dr. Dan Scott says, "The cure of any disease or injury is essentially a process of adaptation." A colleague, Dr. Bach, replies, "You were proceeding under the theory that all living organisms possess the ability in more or less degree to heal themselves." Scott quickly adds, "By adapting themselves to any harmful change in their environment. A lizard for example when injured will shed a tail and grow a new one. A chameleon will change its color for self protection." Bach says, "And you hope to develop a cure of serum from insects since they are the most adaptable of all living organisms." Scott replies, "So, I have developed a serum from the most highly evolved and most adaptable of all insects. The fruitfly ... I putrified the bodies, injected into a cow, and produced the serum. After clarifying with albumin, evaporating in vacuo...." Therefore, the serum Scott developed is cow based and not human nor insect.

A patient, Kyra (Mari Blanchard), had a life-ending lung infection, tuberculosis. A chest x-ray shows her left lung lobes to be very opaque meaning there was fluid infiltration, typical of TB patients. Dr. Scott gives her a 10cc IV injection saying it will require "6 hours to take effect." After injection, there was no sign of a needle mark since it had "closed and healed simultaneously." Later, a leopard scratched her arm and the wounds quickly healed (in seconds). These results are impressive, but in reality neither the growth hormone nor such a serum would act that quickly.

From the emotional stress of being caught Kyra changes her hair from black to blond to avoid capture. In the film, Kyra can change her hair back to black and then back to blond just by "thinking" about it. Bach later looks at her hair sample under a microscope and comments, "the pigmentation is undoubtedly natural," meaning no dye pigment. Scott says, "It could have been the serum that changed Kyra's hair. Bach comments, "Emotional disturbances cause glandular disturbances. They in turn produce physical changes. Maybe it wasn't the serum alone that changed her hair but the serum and some great emotional disturbance." Either way, all hormone based.

Physiology

The Alligator People (1959)

Growth hormone.

Dr. Mark Sinclair (George Macready) says, "Then you know something of the life processes of the higher and lower organisms. Species like ourselves with a highly developed nervous system, bodily functions are controlled principally by the brain and the nerves. But in creatures with a less complex nervous system development life processes are governed by chemical substances secreted by ductless glands and carried in the bloody system." Joyce responds, "Like hormones." Sinclair continues, "Being a doctor I was tremendously impressed by the healing power of one hormone, hydrocortisone. It occurred to me how much more potent this hormone would be in a simpler organism.... There are some small lizards that when attacked detach their tails completely.... There are one or two species that can replace an entire limb if they lost one.... I wanted to extract this wonderful reptilian substance and use it to cure human injuries. I isolated a protein chemical from the anterior pituitary glands of ... our common variety of alligator."

To stimulate the pituitary gland secretions Sinclair uses radiation therapy via a 30 sec exposure from a cobalt 60 source (note: the radioisotope, cobalt 60, produces gamma radiation). The cobalt beam is aimed at an alligator brain at a 45 degree angle.

Volunteers mangled from auto accidents or with other extreme injuries are injected with the serum, in an attempt to fix or repair human injuries. Sinclair explains, "There was an additional secretion beside the healing hormone (growth hormone) that I didn't know about." Joyce Webster (Beverly Garland) helpfully notes, "Your patients are turning into alligators."

Hydrocortisone is the same substance as cortisol and is a steroid hormone made in the adrenal cortex and is released in response to stress. Its primary role is to increase blood glucose, suppress the immune system (suppresses inflammation), and helps metabolism. A rise in cortisol levels can impact decision making and the more the stress the more cortisol is made. Cholesterol is a metabolic precursor of steroid hormones like testosterone, estrogen, and cortisol and is the main culprit in synthesizing cortisol.

The Giant Gila Monster (1959)

Growth hormone.

Many myths, legends, and folklore surround the Southwest and near the top of that list is the Gila Monster. In this film, the giant Gila monster is made possibly by radiation induced pituitary hormone production. The sheriff (Fred Graham) says, "I've been talking to a zoologist and the Gila monster's size is controlled, like everything else, by a sort of thyroid or pituitary gland. Sometimes a change in diet can throw the balance all out of whack. Either the cells break down too fast or build up too slow and this makes runts or giants out of them." Then by way of explanation the

10. Hormones, the Scariest of Them All!

sheriff says, "Certain salts have been absorbed by the plants, eaten, and transferred to animals causing them to be giants." (Why weren't the plants affected?)

The Gila monster (*Heloderma suspectum*) is the only venomous lizard native to the U.S. With "monster" as part of its name it inevitably has a tough reputation. It is easy to identify with its bright pink, orange, or red markings against a jet-black bumpy or beaded skin. And though the giant Gila monster in the film wreaks havoc and destruction, gilas, monsters or not, aren't all bad. In reality, the Gila monster is a shy and slow moving nocturnal animal that still inhabits the Gila River valley in Arizona and New Mexico. Legend has it that the Gila monster has toxic breath and that its bite is fatal (which it is not). It does not have hollow fangs (like venomous snakes) but rather has grooves in its lower teeth that channel the venom. This is why the Gila monster bites down on its victims and this chewing action works the venom into the bite.

An interesting side note is that Gila monster venom has been extensively studied and one particular component is being used to treat type 2 diabetes.

The Leech Woman (1959)

Dr. Paul Talbot (Phillip Terry), an endocrinologist, journeys to Africa in search for a hormone that restores beauty and delays aging. Talbot gets a blood sample from a woman (Estelle Hemsley), claiming to be 152 years old. He looks at the sample under a microscope and confirms her claim. But based on the woman's elderly appearance he concludes, "Nothing can reverse the aging process." She counsels him that restoring a youthful appearance requires, "another substance mixed with nipae powder." To demonstrate she places a few pinches of the powder into a glass and drinks and she appears younger. Since the mixture is ingested natural digestive processes do not affect the formula.

Nipae, a powder from an orchid, is one active ingredient of the anti-aging formula. The second active ingredient is fluid from the pineal gland, which when paired with the powder has a two-fold effect of postponing death and restoring youthful beauty.

The pineal fluid is obtained by plunging a two-inch-long lance into the lower neck area of the chosen victim. It should be noted that the pineal gland is in the center of the brain about four inches deep. It is difficult to understand how pineal hormones were obtained in this manner. June Talbot (Coleen Gray) learns that she is in Africa merely to serve as a test subject and chooses to sacrifice her husband Dr. Paul for pineal fluid. The hormone effects are temporary and a native says to June, "Your youth will not last long. Enjoy it." June changes back into an older looking woman (significantly older each time she reverts, after the nipae wears off) as the stress of the hormones wearing off ironically accelerate the aging process. She does not hesitate to sacrifice more males to prolong the effect, until the accelerated aging leads to its inevitable conclusion.

Physiology

Atom Age Vampire (1963)

Growth hormone.

Professor Alberto Levin (Alberto Lupo) develops a compound, "Derma 28," which regenerates and rebuilds abnormal cells and tissues. The motivation behind this is Levins' desire to help those suffering from radiation-induced burns and tissue damage. As Levins says, "the destructive and degenerative effect of atomic explosions have driven scientists more than ever before into research involving methods and processes of regeneration, rebuilding abnormal or totally destroyed cells."

After using all his supply Levins kills young women to get the "glands which produced derma 28." Most likely, these glands are the pituitary, hypothalamus, and possibly the adrenals. (In this film, a young female is interpreted as post-puberty and pre-menopausal where "glands" would be more potent.) Nevertheless, a key ingredient would be estrogen with the supposition that female hormones would be better at returning scarred tissue to normal tissue in females. All in all, the radiation scarred cells and tissues would be destroyed and replaced with new cells and tissues brought about by the hormones and this would take some time to occur.

Corruption (1968)

Estrogen.

Lynn Nolan (Sue Lloyd), the model girlfriend of physician Sir John Rowan (Peter Cushing), is injured when a hot flood lamp falls on her face badly scarring her right cheek. Rowan's research on guinea pigs showed that "living tissue can be restored without the pain of continuous grafting." During an autopsy Rowan removes the golf-ball-sized "pituitary gland" from a woman killed in a car crash. As Rowan comments, "I've taken the pituitary. I believe I've discovered an entirely new way to control the endocrine system to promote tissue growth." He then makes an extract and injects about 2ml directly into the scar tissue on his girlfriend's cheek saying, "This contains the fluid that controls growth through the endocrine system." Unfortunately, the scar tissue returns about two weeks later demonstrating the temporary nature of the hormone treatment, and causing Rowan to kill three more females to get their glands to treat his girlfriend.

Conclusions

The films overall can be categorized by the types of hormones used: growth hormone, estrogen, testosterone, nerve growth factor, and endorphins. Each hormone is specific to what physiology is being regulated; growth hormone for size (gigantism); estrogen for youth and beauty; testosterone for virility and strength; nerve growth factor for brain effects; endorphins for euphoria. All in all only five different human

10. Hormones, the Scariest of Them All!

hormones and one plant hormone are used in these SF films suggesting that the extensive list of possible hormonal effects has just barely been tapped.

Growth hormone used to alter appearance / Excessive growth hormone
Dr. Jekyll & Mr. Hyde
Island of Lost Souls
Dr. Renault's Secret
The Monster Maker
The Mad Monster

Radiation inducing gigantism via growth hormone from the pituitary
The Amazing Colossal Man
The Beginning of the End
Mesa of Lost Women
Giant Gila Monster
Night of the Lepus (though not discussed in this chapter the film plot uses hormones to disrupt rabbit reproduction, but instead, created giant [and very hungry] rabbits)

Pineal gland
She-Devil
The Leech Woman

Plant hormones
The Beginning of the End

Making Women More Beautiful
There are many films about transforming female scar tissue back to original form and this is a selective list. Over a 25 year period (1942 to 1968) the same plot device of making a woman beautiful with hormones has continued to be mined.

The Corpse Vanishes (1942)
She-Devil
The Leech Woman
Atom-Age Vampire
Corruption (1968)

Summary

Hormones are the scariest of all biomolecules because in one way or another they control *all* of our body's responses, both good and bad. They do indeed determine everything. That being said, then virtually every SF film, not to mention many outside the genre (including westerns, romance, and certainly testosterone-filled action films), has some sort of hormonal influence, either directly or indirectly. This means there

Physiology

are too many films in which hormones play a role so we have to be selective and those selected here are good representative examples. The films discussed here historically began during the gland phase of cinema (early 1930s) and progressed into the hormone phase which in essence is still with us. This could be called the "hormone gland era" in SF cinema. Since then the world has moved from one of hormones, a transient control system, to a more fundamental control, namely that which controls all life, namely DNA. And when you control DNA then you effectively control life and all its forms. In the 21st century the plots have shifted away from hormones and directly to the source of life itself, DNA.

11

Invasion of the Microbes

Like it or not microbes are everywhere and the good news is just about all of them are our friends. It's the few enemy microbes that cause all of the problems. In hindsight it is surprising to find so many SF films that have microbes in their story lines. This reflects the amazing diversity of microbes and helps show how really prevalent they are in our lives.

According to Joshua Lederberg, Nobel Laureate, "The single biggest threat to man's continued dominance on the planet is the virus." Worldwide infections now seem like a monthly occurrence. The "Next Big One," bird flu, swine flu, super flu, hemorrhagic fever, plague, black death, SARS, and MRSA make up quite a scary list. These invisible invaders do bring about thoughts of a doomsday. The World Health Organization and the Centers for Disease Control issue regular warnings of outbreaks, epidemics and pandemics, instilling fear of horrible diseases leading to a possible mass death. Since cinema does shape and influence society then depictions of fictional microbe invasion scenarios do influence the public's perception of what real-life microbes can do.

Microbe Size

Microbes are tiny creatures too small to be seen with the naked eye and can either be a single cell or a multicellular organism. Microbes are so tiny that millions can fit into the eye of a needle. How tiny is tiny? Well, it depends upon scale. One microbe's Lilliput is another's Brobdingnagian. Bacteria, fungus, archaea, parasites, yeast, spores and protists are all considered microbes. Archaea are bacteria-like microbes that have some traits that are unique and not found in true bacteria. Protists include algae, amoebas, slime molds, and protozoa. Though microbes are living creatures in that they self-replicate we can also include viruses as microbes even though they, as a class, do not self replicate.

Microbe Habitats and Ecology

Microbes are found everywhere in nature and represent the majority of life on Earth by populating a wide range of niches. They live in every nook and cranny of

Physiology

the planet including soil, hot springs, the ocean floor (microbes have been reported inhabiting the Mariana Trench, the deepest part of the ocean), ice flows, and the atmosphere. Our frozen poles, deserts, and rocks are all populated by microbes. The amount of living organisms below the Earth's surface is comparable to the amount of life on the surface. Some microbes can survive prolonged exposure to vacuums and can be resistant to radiation (suggesting they can survive in space; scary indeed). Many microbes have a symbiotic relationship with their hosts which are mutually beneficial whereas others can cause disease. Microbes that cause disease are called pathogens.

Microbes are the oldest forms of life. They have been found preserved in 220 million year old amber, which reveals that their size and structures have remained relatively unchanged in all those millennia. And microbe fossils date back more than 3.5 billion years to a time when the Earth was covered with oceans that regularly reached the boiling point, hundreds of millions of years before dinosaurs roamed the earth.

Microbes play a vital role in nutrient recycling in a variety of ecosystems. They decompose matter into more useable forms including fixing nitrogen as part of the plant nitrogen cycle. Plant roots contain microbes that fix atmospheric nitrogen making this important molecule biologically available for plants to use. In this way microbes have developed a symbiotic relationship with plants. And as airborne particles they may even play a role in weather and precipitation. Also, microbes "invented" photosynthesis by having oxygen as waste product and currently make about half of the oxygen on the planet each year and indirectly for most of the rest. Ancient cyanobacteria became chloroplasts, the photosynthetic oxygen producing engine of plant cells.

Microbes were first discovered by Anton van Leeuwenhoek in 1675. Using the microscope he created, he showed that there were forms of life that were not visible to the naked eye. Until the microscope was invented microbes were unknown. Microbes are currently being used in the biotechnology industry from fermentation and beverage preparation to making new medicines through genetic engineering.

Types of Microbes

Microbes do not simply swim—they push and pull, twitch and skitter, spin and corkscrew their way through fluids and across slimy surfaces. Microbes have evolved sophisticated and unusual ways of moving on their own. Many microbes have tiny appendages, called cilia and flagella that help them move about whereas others use intricate protein motors and slime-thinning enzymes.

Most animals are multicellular but there are a few microscopic arthropods such as mites and some nematodes (worms) that are quite tiny. Another example are the rotifers which are filter feeders found in fresh water.

11. Microbes in Science Fiction Cinema

Fungi have several unicellular species the most common being *Saccharomyces cerevisiae* (commonly known as baker's yeast). Some forms of yeast such as *Candida albicans* are pathogenic to man.

Since microbial algae, green algae, contain chlorophyll and are photosynthetic they are considered plants. These forms of algae can grow as single cells or as long chains of cells. In total there are about 6000 species of green algae.

Extremeophiles are microbes that have adapted to such extreme conditions that are fatal to most life forms. Such microbes have been found in temperatures as high as 130°C (266°F) and as low as −17°C (1°F); in acidic environments with a pH of 0 and alkaline environments with a pH of 11; from high pressures of 2000 atmospheres down to 0 atmospheres (vacuum of space); and exposed to mega doses of radiation (up to 5000Gy). Much of this extreme adaptability has been exploited by the biotechnology industry in generating unusual products.

Bacteria

Cell organisms that lack a cell nucleus are called bacteria, or prokaryotes, and are almost always unicellular. Almost all are invisible to the naked eye. Their genome is just a single loop of DNA and bacteria have a sturdy cell wall, providing strength and rigidity, that protects and isolates its internal cellular components from the environment. In total, prokaryotes are the most abundant and diverse group of organisms on Earth. They inhabit all environments that are below +140° C and are found in water, soil, and air. They are found in such diverse places as hot springs, deep beneath the Earth's crust, and gastrointestinal tracts. As a mechanism of survival some species of bacteria form extraordinarily resilient spores that can withstand extreme environmental challenges. And under optimal growth conditions bacteria can double in as fast as 20 minutes. It has been estimated that there are around five million trillion trillion (that's a massive 5×10^{30}!) prokaryotes accounting for at least half of the Earth's biomass.

Pathogenic Bacteria

Symptoms of bacterial infections are localized redness, heat, swelling, and pain. In particular, localized pain is a key characteristic of such infections. For example, when a cut is infected the pain is localized at the site. For bacterial throat infections the pain is localized to one side of the throat whereas for a bacterial infection in the ear the pain is localized in only one ear. Cuts that produce pus or a milky liquid are most likely due to bacterial infections. Bacteria microbes release various toxins that can easily hobble host cells, oftentimes killing them. One recent example of this is toxic shock syndrome caused by bacterial toxins.

Physiology

Viruses

There are thousands of types of viruses and they are everywhere. There are an estimated 10^{31} viruses on earth! What this means is there may be a hundred million times more viruses on earth that there are stars in the universe. A liter of seawater near the surface contains 10 billion microbes and 100 billion viruses. Also, special viruses that attack and control bacteria, called bacteriophages, are replicated at an astounding rate of 2.5×10^{25} viral genomes every second.

Most viruses are composed of a tiny piece of DNA surrounded by a protein coat and sometimes an outer envelope. Though viruses have genes, they cannot replicate but must hijack the DNA of a living cell. They bind to the outer surface of a host's cell and disrupt the cell's own genetic machinery into making more virus particles which in turn go on to infect other cells. Viruses as a class cannot convert carbohydrates into energy so they do not metabolize and must rely on the host cell to do that.

In some instances the virus kills its host cell which can lead to certain diseases. As a class, viruses are very tiny and are much smaller than bacteria. Such common diseases as the cold (rhinovirus), flu (influenza virus), and cold sores (herpes virus) are all caused by viruses. Also, such deadly diseases as HIV/AIDS, smallpox, and some cancers are caused by viruses.

Since viruses "live" inside cells they are difficult to treat without harming the cell, but effective vaccines have been developed that immunologically treat viral infections. Due to the physical nature of viruses antibacterial medicines are not effective, but the body's natural defense system, the immune system, is capable of killing most viruses. Often people do not know they were even infected as the natural immune system took care of the virus before any symptoms were present.

Pathogenic Viruses

Most viral infections are systemic meaning they infect many parts of our bodies at the same time. Even though the infection may be localized (like a nasal cold, "pink eye" or viral conjunctivitis) few are actually painful, like herpes. Since many of these viruses also attack nerve cells many symptoms of viral infections are described as a burning or itchy feeling. Some pathogenic viral infections can cause cancer.

Parasites

Organisms that live off other organisms are called parasites. Overall, there are two types, ectoparasites that live outside a host and endoparasites that live inside a host. Obligatory parasites require a host throughout their life cycle. There are microparasites such as viruses and bacteria and macroparasites such as mites and fleas. Human parasites include protozoa (such as malaria, algae, and amoebas), worms (tapeworms), and flukes that can either cause infections inside the body or topically on the skin surface. Other examples of ectoparasites are lice and ticks. Also, parasites can be used as vectors to infect various forms of DNA into its host. Typhus, a deadly flu-like disease, is spread by body lice. There are subtypes that infest the head and others that infest the body which thrive in clothing. Malaria carrying mosquitoes also use parasites as a vector to carry on the infection.

Fungus

Mycology is the study of fungi and yeast. A fungus is a eukaryotic, heterotopic, spore-bearing organism with chitinized cell walls. Fungi are common in the environment and are found in soil, on plants, trees, vegetation, skin, mucous membranes, and perhaps somewhat surprisingly, commonly in the human intestinal tract. Though most fungi are dangerous to humans some can be beneficial; penicillin, bread, wine, and beer use fungi in their manufacture. Mushrooms are also a type of fungus.

Some fungal infections are mild resulting in a rash whereas others are very pathogenic and can lead to serious complications like meningitis. Those with a weakened immune system, such as cancer patients, transplant recipients, and those with HIV/AIDS are susceptible to opportunistic fungal infections. Also, climate change could affect some fungi since even small changes in temperature or moisture can affect their growth (see *House of Dracula*). The potato famine in Ireland during the late 1840s was due to a fungal infection that wiped out the crop.

Microbe Fun Fact: No Man Is an Island

Humans are made up of about 10 trillion cells and also harbor about 100 trillion microbes so by cell count about 10 percent of the total cells of our body are human. Since microbes are much smaller than human cells the 100 trillion microbe cells weigh about 3 pounds, about as much as a human brain. Though there are roughly 21,000 genes that makes us human, microbes add another 8 million or so genes, many of which handle behind-the-scenes activities such as processing food, tweaking the immune system, and helping to regulate some human genes by turning them on or off. Given all of this microbe activity, no man is an island but rather a metropolis. So the total human genome (hologenome) is a combination of human DNA *plus* microbe DNA. Ultimately, microbes help establish a healthy genetic equilibrium.

Microbes inhabit almost every corner of the body and all told there are more than 10,000 species in and on our bodies. For example, there are over 1000 microbe species in your mouth; 150 species behind your ears; 440 species on the insides of your forearm; 5,000–35,000 bacterial species in your intestines. Even microbes on our right and left hands are different and they have only about 17 percent of species in common between the hands. In total, there are more microbes on a person's hand than there are people on the entire planet! Of all these thousands of species of bacterial microbes only about 100 are pathogenic to man. Interesting to note that 33 percent of humans (with or without illness) are colonized with *Mycobacterium tuberculosis*; 50 percent of humans (with or without illness) are colonized with *Helicobacter pylori*; and 50 percent of humans (with or without illness) are colonized by *Staphylococcus aureus*.

Of the 1000 species of bacteria that live in the human mouth, some are harmful (those causing cavities) and some helpful (keep breath smelling fresh) but most are unspecified. In total, more than a trillion microbes live within your mouth and they multiply so fast that you literally swallow trillions a day without even denting the population. Furthermore, in total, millions of bacteria are enjoying a comfortable stay on your face. The rest of the body is no better with your guts being, so to speak, the mother lode.

Animals, including humans, couldn't digest food without them. Also, without microbes, plants couldn't grow, garbage wouldn't decay and there would be a lot less oxygen to breathe. Microbes also help in remediation and recycling of nutrients. So, microbes are essential to the processes of decomposition that recycles nitrogen and other elements back into the natural world. At the human level, disruption of our microbiome can lead to certain disorders such as obesity, allergies, diabetes, bowel disorders, even psychiatric disorders like autism, schizophrenia, and depression.

Gut Microbiome

The human gut microbiome contributes 36 percent of the small molecules that are found in human blood. On the down side some gut microbes contribute greatly in creating susceptibilities to certain human diseases. The human gastrointestinal tract (GI), which includes the stomach and intestines, harbors trillions of bacteria belonging to more than 1000 species and there are 10 times as many bacterial cells within the GI tract as there are human cells within our bodies. The GI microbiota plays essential roles in human nutrition, physiology, development, immunity, and behavior so disrupting the structure and balance of these microbes can lead to disease. Most of these microbes are our friends, helping us digest food, strengthen our immune systems, and keep dangerous enemy pathogens from invading our tissues and organs. Many people, especially relatives and family members, share gut microbe strains most likely through touching each other or sharing the same environment.

There are good microbes (such as those which provide nutrients and help metabolism) and there are bad microbes, those that cause health problems (including those that invade us). This mass of microbes can be considered as an "organ" because they

11. Microbes in Science Fiction Cinema

carry on all the functions of what we consider an organ. After all, if the immune system is considered the "liquid organ" then the combined mass of microbes that have their own DNA, carry out key metabolic steps, and make all sorts of useful biochemicals, should also be thought of as an organ. And as an organ it too needs to be cared for, nurtured, and "fed properly."

How are gut microbes so stable? The gut is continually being washed but these microbes manage to withstand all our bodies can throw at them. They are exposed to severe environmental extremes not to mention all of the natural metabolic and immune responses of their hosts but they still manage to survive. Which is why it is so impressive that these "small soldiers" are so effective an army against alien invaders. They have developed impressive survival tactics that serve us well against invaders of all types, including actual aliens, whose natural bodily defenses are not equipped to handle such "pests" (see *The War of the Worlds*).

Other Uses of Microbes

- Digestion. The natural gut microbes help in synthesizing such vitamins as folic acid and biotin as well as helping to digest complex carbohydrates. These microbes also contribute to gut immunity.
- Food and Beverages. Microbes are used during the fermentation process in brewing beer and wine making. Baking and pickling processes require microbes as well, as does the conversion of dairy into cheese and yogurt. Some microbes add flavor and aroma to the products.
- Water treatment. Sewage treatment is an oxidative process that requires microbes to perform the task. These microbes process sludge that produces, among other things, methane gas.
- Energy. Microbes used in fermentation can also be used to produce ethanol, a form of biogas. Both algae and bacteria have been used for this process to create useable fuels from agricultural and urban waste.
- Production of chemicals. Microbes are used routinely for commercial and industrial applications such as in the making of some forms of biological acids made from bacteria like acetic and lactic acids as well as bioactive molecules like enzymes. An example is streptokinase that is used to dissolve blood clots and some statins used to lower cholesterol levels. Microbes have become essential tools in the biotechnology and biomedical areas since they grow rapidly and are easily manipulated. Microbes have also been effective in cleaning oil spills and used as living fuel cells.
- Warfare. Biological warfare has been in the news recently but the idea has its roots in the Middle Ages when diseased corpses were thrown into castles during sieges using catapults. Those near the microbe-contaminated corpses were likely to spread the deadly pathogen to others.

Physiology

Diseases Caused by Microbes

Examples of some common diseases caused by microbes are AIDS, HIV, bronchitis, cancer, chickenpox, colds, dengue, Ebola, encephalitis, genital warts, German measles (rubella) hantavirus, hepatitis, herpes, influenza, Lassa fever, leukemia, measles, meningitis, mononucleosis, mumps, plague, polio, rabies, shingles, STDs, and warts. Such diseases as plague, tuberculosis, and anthrax are caused by pathogenic bacteria. Protozoas cause such diseases as malaria, sleeping sickness, and toxoplasmosis. Fungi cause such diseases as ringworm and candidiasis. Pathogenic viruses cause such diseases as yellow fever, and, as mentioned, influenza and AIDS.

Viral and Microbial Infections

How do viruses/microbes infect the body? Viruses primarily enter the body through any of the openings, primarily the nose and mouth. Once inside the virus attaches itself to an appropriate host cell and begins to replicate or make copies of itself. The rhinovirus, the one that causes the common cold, finds host cells in the nose whereas an enterovirus, one that causes intestinal issues, finds host cells in the gut. After making copies of itself the virus exits the host cell to find another host cell and begin the cycle all over again. In many cases the virus ends up destroying the host cell. Also, not all viruses attack one part of the body (localized) but some are disseminated throughout the entire body via the blood stream such as the AIDS virus. Latent viruses are infections that are dormant or hidden that are not involved in active signs and symptoms of a disease. Also, some microbes do not cause an illness by themselves, but rather, by the body's reaction to it. If some key tissue, say nerve cells, are viral infected and the body's reaction is to destroy the cells and tissues then some neurological disorders can result.

Viruses and microbes cause illnesses by destroying or interfering with cell functions, especially important cells that are key regulators of body physiology. The virus can destroy a cell by a process called "programmed cell death" (or apoptosis) or keep it from making energy to grow and live by interfering with cellular biochemical balance.

In most instances the viral infection is cleared by the immune system within a few days to a few weeks. Some viruses are persistent and can last years in the host. These viruses can lead to latent infections in which symptoms can appear much later. The infected person may seemingly recover and be normal with the virus still present but perhaps dormant waiting to be re-activated. Examples of these viruses are the herpes viruses, hepatitis B and C viruses as well as HIV.

Microbial disease is usually the result of the host's insufficient immune system. Microbes cause damage by releasing a variety of toxins or destructive enzymes that destroy a host's cells. Some microbes, like *Clostridium tetani* releases toxins that paralyzes muscle cells whereas *Staphylococcus aureus* releases toxins that produce shock and sepsis.

11. Microbes in Science Fiction Cinema

Infectious Disease

An infection is an invasion followed by a multiplication of some sort of parasitic disease-causing organism (a virus or a bacterium can also be considered a parasite) within a host's body. Even larger organisms like macroparasites and fungi can cause infections. These organisms typically enter, invade, or inhabit another body, causing infection and/or contamination oftentimes releasing toxins that can be detrimental to the host. Some infections take over the entire body, and others affect a specific organ, like the brain, lungs, or liver. The world's most deadliest infectious disease is tuberculosis. Approximately one third of all people on Earth are infected with tuberculosis and about three million die every year from this disease.

Infections can either be acute (short-term) or chronic (long-term). Bacterial infections are based on the symptoms and medical signs shown by the patient. Some infections are not readily apparent and referred to as silent or subclinical and others that are inactive, though still in its host, are dormant and called a latent infection. Also, infections can be either primary or secondary depending upon the stage of infection and whether it is the same infection or a succeeding infection in which the offending microbe can repopulate the host or reappear from latent hiding. Furthermore, occult infections are those that are essentially hidden and recognized by secondary symptoms.

Some microbes cause persistent or chronic infections because the body is unable to naturally clear the organism after initial infection. The microbe is continually present in these hosts in amounts low enough to not be a problem. Sometimes relapses occur in which the microbe repopulates giving rise to another infection. Latent infections are also persistent infections that can repopulate by different mechanisms and an example is herpes virus that likes to hide in nerves and when certain circumstances occur become reactivated (cold sores are examples of this).

Each host has its own specific response to infections though most often this response comes from the immune system, both innate and induced. In many instances the invading microbe causes inflammation which activates and stimulates an immune response. Should the host not have an effective physiological means to stop the invading microbe then eventually the microbe will win and could ultimately cause the death of its host.

An infectious disease can cause mass destruction in populations not previously exposed to that particular illness. The Native American Indians were decimated by "European diseases" such as small pox, venereal diseases, and cholera, to which they had no natural defenses or resistance. Smallpox epidemics of 1818 and 1839–1840 and the cholera epidemic of 1849 wiped out large populations of Indians.

Plague

A plague is any disease of wide prevalence that results in excessive mortality or death. Mankind has been affected by various plagues since recorded history, many of which are described in the Old Testament of the Bible. Moses inflicted the plague on Pharaoh Ramses II before the Exodus. In the Middle Ages plagues decimated Europe after the bacterial microbe, *Yersinia pestis*, was transmitted to man from the fleas of rodents. Clinically, signs of plague are high fever, toxemia, prostration, hemorrhagic ruptures, lymph node enlargement, and pneumonia. In man, plague comes in four different forms: bubonic (the most common, marked by extensive inflammation of internal tissues and glands), septicemic (a generally fatal form with high levels of microbe growth and excessive toxemia), pneumonic (a progressive and generally fatal form with excessive fluid build-up in the lungs making it hard to breathe), and ambulant (a mild form of bubonic plague with mild fever and tissue swelling).

Chain of Events

For a microbe to cause an infection a series of events must occur in a particular sequence. First, of course, is the presence of the infectious agent or microbe, then a reservoir to maintain its population, the ability to enter a susceptible host, an exit, and subsequent transmission to new hosts to continue the propagation. Each of these steps must occur in this chronological order for a microbial infection to develop. Should any one of these steps be blocked or inhibited then the infection can be stopped.

Signs and Symptoms

Bacterial and viral infections can both cause the same sorts of signs and symptoms sometimes making a diagnosis difficult. It is important to distinguish between the two since medications for one do not work on another. Symptoms such as fatigue, loss of appetite, weight loss, fever, chills, and aches and pains are common between bacterial and viral infections. Other signs and symptoms can be limited to certain body parts such as skin rashes, coughing and runny noses.

Treatment and Prevention of Microbial Infection

Effective ways to treat and prevent microbe infections rely on disrupting their infection cycle and the easiest way to do this is to maintain a sanitary environment that includes health education. Effective hygiene can be as simple as washing your hands (still the most effective) and for health care workers should also include wearing appropriate gowns and facemasks to help prevent the microbe from being passed from patient to health care workers.

The transmission of microbes is common in hospital settings. Even with careful healthcare practices microbes can cause infections not only in fellow patients but also in healthcare workers.

When microbes do cause infections then the first line of defense to suppress the infection are anti-infective drugs and there are four types: antibacterial, antiviral, anti-tubercular, and antifungal. Drugs that work on one microbe type will not work on another. Depending upon the severity these drugs can be given orally, topically, or by injection. In some cases, cocktails or combinations of drugs are necessary for multiple microbe infections. For bacterial infections the antibacterial drugs act by primarily slowing down their multiplication. The most common antibiotics are penicillin and tetracycline. It is important to not take antibiotics longer than needed to help prevent mutations and therefore resistance to the drugs. Excessive use of antibiotics results in resistant strains like MRSA.

Transmission

For a microbe to repeat its infection cycle it must leave one host and infect another to repeat the cycle again and again. This is how diseases are spread. The transmission from one host to another can take a number of different paths, either by direct or indirect contact. Direct contact includes body fluids, drinking contaminated water, inhalation (including particles from sneezes or coughing), or being bitten by a vector such as a mosquito. Sexual activity also contributes to direct contact transmission.

Indirect contact occurs when the microbe is away from a host and able to infect another host when the opportunity arises. In many cases the microbe is on some sort of object (toys, furniture, hand rails, scary monster magazines) from an infected person waiting for an appropriate new host to repeat its infection cycle. Also, foods can be an indirect source of microbes for disease transmission. In many underdeveloped countries contaminated sewage water is a major source of infectious microbes.

All types of microbes can be transmitted from host to host in this manner and the above examples are of horizontal transmission since the microbe is transmitted from person to person. Microbes transmitted vertically are primarily from mother to child during pregnancy that includes the birthing process; most of these vertical microbes are viruses.

Colonization

An infection begins when the microbe is able to successfully colonize a host by growing and multiplying. Fortunately, most healthy humans are resistant to infections but those who are sick or malnourished have an increased susceptibility to infections. Also, those who have a suppressed immune system are susceptible to opportunistic microbes.

Physiology

The microbe typically enters a body through mucous membranes such as the mouth, nose, eyes, genitals, and open wounds. The microbes often migrate through the body causing general infections that can affect different internal organs. Some microbes grow within host cells whereas others grow in body fluids. It should be noted that some colonizations are not intrinsically infectious. Therefore, all infections are a colonization but not all colonizations are infectious. Some non-pathogenic microbes can, under certain conditions, become pathogenic. The variables involved include the route of entry, the particular virulence of the microbe (more virulent in some species than in others), the amount of infecting microbes, and the overall immune status of the host.

The Films

There are too many SF films that have microbes function as a part of the plot, so we must limit the discussion to representative examples of the microbial world: bacteria, viruses, and parasites.

Son of Frankenstein (1939)

During a medical examination Dr. Frankenstein (Basil Rathbone), looks at a blood sample from the monster under a microscope, observing not only the occasional red blood cell floating by but also a myriad of small moving circles, far more than blood cells. These circles are bacterial microbes, which means this blood sample is heavily contaminated since a normal healthy blood sample would not have *any* microbes. Judging by how the circles are moving in different patterns, it is apparent that there are many different species of microbes present.

The microbes in the monster's blood sample provide two possible scientific interpretations. Either the blood sample was contaminated after it was taken from the monster or the monster's blood actually does contain bacteria. If the latter is true, then we have to explain what the bacteria are doing there and what benefit, if any, they provide for the monster.

Microbes are critical for natural human metabolism and perhaps the monster, stitched together from less than sterile sources (cadavers would be heavily contaminated), has developed a "natural" microbe flora that would be fatal for normal humans. The monster has so much natural microbe flora that it naturally spilled into his blood stream.

Since this film is the monster's third outing, after the original *Frankenstein* (1931) and *Bride of Frankenstein* (1935), along the way his immune system and metabolic needs could have adapted to accommodate a larger number of microbes.

Bacteria secrete endotoxins that damage and kill cells and it doesn't take much to kill a human. With that many bacterial microbes coursing through the monster's blood then large amounts of endotoxins must be released. These endotoxins cause inflammation which can lead to toxic shock and ultimately death but since the monster can withstand such a microbe assault then the shock may actually invigorate his metabolism and provide much energy and super-strength.

House of Dracula (1945)

Baron Latos (disguised as Count Dracula, disguised as John Carradine) sought out Dr. Edelman to "effect a cure" and rid him of his vampire curse. While examining a blood sample from Latos under a microscope Edelman says, "The examination of your blood reveals the presence of a peculiar parasite, the form of which I am completely unfamiliar. It's possible it may have something to do with your problem.... I am having an anti-toxin prepared so that we may see." Later, Edelman says, "a pure culture of a parasite introduced into the parent bloodstream will destroy not only its own kind but themselves as well." In theory, this is what vaccines do.

The bizarre looking parasites are wrapped around individual red blood cells and composed of a long thin arm that branches out to finger-like projections (some have three whereas others have four). Though no such parasite exists there are many that specifically do attack red blood cells, the most serious being the plasmodium sporozoite of malaria.

In trying to treat the parasite Edelman gives Latos a transfusion of his own blood reasoning that the natural antibodies in the doctor's blood would counteract and perhaps destroy the parasites in Latos' blood. This may be one of the first examples of what is now called immunotherapy in that the power and intelligence of the natural immune system is used to treat and regulate a disease.

Once Edelman recognized the existence of the parasite he asks his assistant to "make a culture of this and prepare an antiserum as soon as possible." Well, only in SF cinema can an antiserum be so simply whipped up, since culturing parasites is tricky business that often have unique and unusual conditions for proper growth. An "attenuated parasite" would be achieved through infecting a mammal. And which species of mammal was selected? Typical antiserum species are goats, mice, and rabbits (try to imagine a vampire rabbit). Then there would have to be multiple immunizations with parasite extracts which are typically spread over several months.

Let us take a closer look at the parasite seen in Latos' blood since we get a good point of view shot of it. Also, an important question to ask is what contribution did this organism make to vampirism? Latos, as a vampire needs blood to survive. If the parasite feeds off red blood cells then over time the number of RBCs would diminish resulting in anemic-like conditions and a "transfusion" would be beneficial in helping to alleviate the symptoms. And like many infections the parasite population most likely rose and fell depending upon a variety of conditions. If the parasite population were high then severe anemic conditions could result making Latos have a 'craving' for more blood. The converse is also true when the parasite population is low then there would not be any anemic conditions and the craving for blood would be lessened. Since vampires apparently need blood nightly then the parasites must have rapid doubling times of less than 24 hours.

The War of the Worlds (1953)

After the military proves ineffective in dealing with the Martian invasion, scientists are consulted. Major General Mann (Les Tremayne) explains, "Guns, tanks, bombs (including atomic), they're like toys against them ... our best hope lies in what you people (scientists) can develop to help us." This is one of the earliest films that uses germ warfare as part of its plot.

Dr. Clayton Forrester (Gene Barry) says, "We know now we can't beat their machines. We've got to beat them," further noting, "If they're mortal, they must have mortal weaknesses." While examining a Martian blood sample under a microscope another scientist says, "I don't ever remember seeing blood crystals as anemic as these. They may be mental giants, but by our standards, physically, they must be very primitive."

A spaceship lands near a farmhouse and the Martians come out, wearing no spacesuits nor protective gear, to investigate. And this sets the stage for another invasion to begin, namely that of the immunologically vulnerable Martians by the Earth's microbes, which they are immediately exposed to in the air and through the ground and everything they touch. We see a Martian touching Sylvia's (Ann Robinson) clothing in the farmhouse, another, more intimate, opportunity for infection to set in. And once one Martian has been infected they will transmit the microbe germ to other Martians, kicking off the inevitable chain of events of the infection cycle. The converse could also be true in that humans touched a "wet" scarf stained with Martian blood so this begs the question, do the Martians have their own microbes (likely) and were any transferred to humans via the bloody scarf?

With the death of the Martians bringing the invasion to an end, a voiceover intones, "The Martians had no resistance to the bacteria in the atmosphere to which we have long since become immune. Once they had to breathe our air, germs which no longer affect us, began to kill them. The end came swiftly. After all that men could do had failed the Martians were destroyed and humanity was saved by the littlest things which God in His wisdom had put upon this earth."

Near the end a Martian arm is seen falling out of a hatch that dehydrates and turns a whitish color. Forrester touches the arm of the deceased Martian, potentially transferring Martian microbe germs to humans, raising the odds of a microbial Martian invasion, unleashed onto the immunologically vulnerable Earthlings.

Beast from 20,000 Fathoms (1953)

After atomic blasts awaken a monstrous beast in the frozen north, it seeks out its ancient breeding grounds in New York harbor. The wounded creature roams the

11. Microbes in Science Fiction Cinema

From *The War of the Worlds*: Dr. Forrester (Gene Barry) touches the Martian arm and all those Martian microbes are now on him and anyone else he touches. Thusly begins an infection.

city trailed by large blood stains on the street. Soldiers walking next to blood stains subsequently become ill and soon after collapse; it takes less than five minutes exposure so the pathogen is fast acting.

A doctor informs Col. Jack Evans (Kenneth Tobey), "The monster is a giant germ carrier of a horrible virulent disease. Contact with the animal's blood can be fatal. If you use shell fire who knows how far the air will spread the particles of it." The colonel responds, "Should have used flame throwers. Would have cremated the beast and the plague with it." But the doctor knows better, "The smoke would have carried the blood particles just as far as the air." Note: if contact with the animal's blood can be fatal then what about other body fluids such as urine? Or even the animal's breath.

Another doctor takes a blood specimen from a patient with a temperature of 105 degrees, and orders, "Get it to the laboratory quickly. I'm afraid ... of what that creature has brought to us. Dead afraid." In this scene the handling of the specimen is less than satisfactory and if the "blood particles" are as virulent as they say, all those in the room would become infected.

A scientist advises, "There is only one way to beat him. Use radioactive isotope. Shoot it into him and destroy all that diseased tissue." He continues, "The only isotope of its kind is this side of Oak Ridge." Oak Ridge National Laboratory (ORNL) is a government facility in Tennessee that produces radioisotopes for the scientific community and furnishes radioactive samples to the East Coast. In my research career I have used radioactive samples obtained from ORNL.

Radioactive isotopes have a long half life, some measured in years, so after the beast was killed that isotope is still in him and just as deadly. Death by isotope, meaning radiation poisoning, is a relatively slow process, so in spite of the creature's wounded state it could take weeks to take effect, and not the brief time portrayed in the film.

The Angry Red Planet (1959)

After landing on Mars one of the crew instructs an on-board scientist to take a microbe count to assess what type of life is present, but the likelihood of humans being able to identify Martian microbial life is low.

An astronaut says, "The atmosphere is pretty much like we thought—thin, extremely thin. Not enough oxygen to sustain us, but undoubtedly enough for some kind of native animal life ... with all that vegetation out there [looking out a window in the rocket ship], there's bound to be something alive." It appears that plant vegetation is not considered "alive," at least on Mars.

In the film a giant amoeba attacks the ship and crew. A scientist comments, "I'm sure it's a unicellular animal. The two areas inside it must be the nucleus and the contractile vacuoles. Like an amoeba, a giant amoeba. One single cell without intelligence, without a nervous system at all. It reacts completely on instinct to external stimuli. The amoeba engulfs its prey and digests it with extremely strong acids." The amoeba is larger than the rocket ship, so it's physically impossible for it to be a single cell as its sheer weight would flatten it! Part of this amoeba comes in contact with the arm of astronaut Col. Thomas O'Bannion (Gerald Mohr) and eats right through his clothing and into his skin, causing an infection that is brought back to Earth with the crew.

After the return landing of the rocket ship, the recovery team wears no protective suits, gloves or masks as they assist the contaminated colonel, showing no concern for the potential spreading of infection.

Doctors examine O'Bannion's infected arm and conclude, "An enzymatic reaction. A minute particle of the amoeba creature must have reached Tom's skin, and it's growing, literally eating his tissues." A scientist then realizes, "We've been attacking the alien amoeba as if it were a disease. But it isn't. It's an animal, an animal with instincts. And the most important of all, a will to act."

Taking advantage of amoeboid instincts the scientists use electricity to force off the "amoeba particles" on Tom's arm thereby saving him. For most infections humans spike a temperature and during Tom's infection he spiked a temperature which supports some sort of microbe attack.

As the doctors are examining the infected arm they discuss the risk of infection. During the conversation one doctor takes off his examination gloves and drops them into a bag a nurse (ungloved and unmasked) is holding, again demonstrating very poor containment conditions for a potentially devastating infectious disease, while simultaneously warning, "Suppose this alien infection spreads to all of us. Every moment counts." End scene.

Even though the astronaut appeared to be cured of the infectious disease it is unknown if some of the alien amoeba DNA was able to integrate into his cells and tissues, much the same way natural viruses work. Over time, as a latent infection, this Martian DNA could eventually take over his body creating a major infectious disease outbreak (much like herpes outbreaks come and go). Who knows what could happen if alien DNA were mixed with human DNA?

The Last Man on Earth (1964)

A plague is unleashed upon the world that turns people into vampire-zombies. Ultimately, those who are exposed to the microbe either die or become like vampires in that they are repulsed by sunlight, mirrors, and garlic and like zombies in that they are driven to consume flesh. Dr. Robert Morgan (Vincent Price) kills the infected with a wooden stake through the heart.

Corpses cover a desolate and otherwise empty urban city. In an effort to clean up, Dr. Morgan collects bodies and takes them to a massive crematorium for disposal and while doing this he wears a gas mask to prevent inhalation of the germs. Apparently, animals are not infected since we see a dog later in the film who appears normal so the microbe appears to be human specific.

In an early discussion between Dr. Morgan and his wife Virginia (Emma Danieli) he says, "I can't accept the idea of a universal disease." His wife responds, "Is it possible this germ or virus could be airborne?" Morgan responds, "The germ is visible under a microscope but is not like any bacillis ever known. It can't be destroyed by any process we've been able to uncover."

Symptoms of infection by the microbe are hard breathing, fever, and cramps with eventual blindness. It should also be noted that the infected also feel pain since a door was slammed on the arm of one who yelled, "ow" in response. Since the wife says "virus" and the husband says, "bacillis" it is confusing. As mentioned above, treatment for one microbe will not work on another. Furthermore, real viruses are too small to be seen using a standard light microscope as seen in this film. To "see" viruses you need an electron microscope.

In a scientific exchange at the Mercer Institute of Chemical Research, Morgan says, "The bacilli are multiplying." A microscope view of the "bacilli" shows the microbes to look suspiciously like spirochetes and definitely NOT bacilli, nor viruses. Then, "And this bacilli is found in the blood of every infected person." Dr. Mercer (Umberto Rao), the head of the institute, responds, "That kicks the bone marrow theory in the head." Morgan then adds, "This specimen shows a higher white count than when I put it on the slide. Those cells are still living off one another." (If they were living off one another then their number would decrease due to eating each other and not "show a higher white count.") Mercer instructs, "You two stay on this virus theory." Not sure what a "virus theory" has to do with the problem of bacillus microbes.

Morgan can't bring himself to slay an infected woman Ruth Collins (Franca Bettoia) who asks how he has survived. Morgan replies, "A long time ago when I worked in Panama I was bitten in my sleep by a bat. My theory is the bat had previously acquired the vampire germ. By the time it had entered my blood it had been strained and weakened by the bat system. As a result, I have immunity."

The bite essentially vaccinated Morgan to the germ, which suggests transfusions of his blood could if not kill the germ then at least contain it. Morgan gives Collins a transfusion and explains, "The blood feeds the germs and the vaccine keeps it isolated. It prevents it from multiplying ... the antibodies in my blood worked. My blood saved you." Later, she says, "We're alive. Infected, yes ... but alive." Human antibodies do indeed work and it is satisfying that this 1964 film got that right.

Flesh Eaters (1964)

Dr. Peter Bartell (Martin Koslec), a professor of marine biology, is on a remote island researching flesh-eating organisms. It should be noted that flesh eating microbes are real and do pose a serious threat. The worst of the bunch is *Staphylococcus aureus* (SARS; staphylococcus aureus resistant strains) that gets into muscle. Those infected have a difficult recovery and often lose limbs and tissues as a result. Bartell, a German scientist, is inspired by some Nazi war experiments where biochemists were accused of producing a life form that had a "peculiar metabolism," requiring only one form of nourishment. In his well-equipped laboratory, housed in a tent, he uses electricity to create a giant parasitic amoeba from the flesh-eating microbes. In explaining this, Bartell says, "The charge of energy bound the amino acids together and it formed this." In essence, Bartell used electricity to congeal the small microbes into a larger one (much like real slime molds congeal into a stalk and bulb). The amoeba has external structures such as appendages, cilia, an "eye," and surface ridges, which an actual amoeba would not have.

A plane is forced to land on the island during a storm and the passengers encounter the effects of Dr. Bartell's experiments. A bone white human skeleton is seen on a beach. Nearby are "glowing" fish skeletons and the ocean water is glittering with illuminated flashes of life. When Grant Murdoch (Byron Sanders) reaches to touch a fish skeleton Bartell warns, "What you are about to do is extremely dangerous. I would guess that these fish were destroyed by some microscopic parasite. There's the possibility this same parasite could be transferred to your body if you should touch them."

Later, Murdoch warns his friends that there is something in the water that "eats the skin right off you." This is soon demonstrated on an unfortunate man on a nearby raft, whose flesh dissolves when ocean water splashes his face. After he drinks from the same water, his internal organs are eaten by parasites from the inside out. The voracious activity of the parasitic amoeba is shown by the man's emerging skeleton with fumes coming off his clothes (a popular SF effect achieved by bits of evaporating dry ice).

Upon discovering that the parasite has "hemoglobin sensitivity" the other survivors from the plane pool their blood to use as a weapon and use a crude hypodermic syringe to fatally dispatch the giant amoeba. The main component of blood is hemoglobin which contains iron atoms so it seems appropriate that iron was the toxic element that killed the parasitic amoeba. The amoeba essentially died of lead poisoning.

Outbreak (1995)

An African monkey carrying a deadly virus is smuggled into the U.S. and people exposed to the virus develop hemorrhagic fever, the main symptoms being loss of blood through blood vessels. Most viruses are confined to one species, but the "Motaba" virus featured in the film is a zoonotic virus, meaning a virus that jumps from one species to another. First the virus infects the African monkey, then it mutates to infects humans, and then an even more deadly mutation results in an airborne virus, the most contagious form, which infects a whole town.

Casey Schuler (Kevin Spacey) describes the gory final stages of the illness, "When the patient first gets the virus he has flu-like symptoms and in two or three days pink lesions begin to appear all over his body, along with small pustules that soon erupt with blood and pus. A kind of milky substance begins to form. These particular lesions, when they become full blown, feel like mush to the touch. There's vomiting, diarrhea, bleeding in the nose, ears, gums, the eyes hemorrhage, the internal organs shut down, and then they liquefy." Examples of hemorrhagic viruses are ebola, lassa, and hanta viruses.

As it turns out the military had developed an antidote that saves the inhabitants of the infected town, and prevents the worldwide spread of death by liquefaction.

Summary

Microbes come in all sorts of shapes and sizes but in essence all share a common trait. They hijack a host's biology to serve their own needs, sometimes to the benefit of the species and sometimes to its detriment, often in the form of a contagious illness.

The eight films discussed here cover a span of over 55 years and during this time infectious diseases have become more prominent in our lives. The possibility of global

Physiology

infection lurks in the depths of our collective psyche and every new outbreak increases that concern. The idea that a foreign visitor, an "alien," can take over our bodies is troubling and accompanied by a sense of helplessness. Of course, historical diseases such as the Black Death and bubonic plague predate film, but the medium effectively captures the angst of the public over the very real, but invisible threat of microbes, a fear that feeds on contagion. Scary indeed.

Surgery

This article was the first in this loose series of "science in SF films" for *Scary Monsters* magazine. As a teacher at UCSD it quickly became apparent that the students were far more receptive to difficult material if presented in an entertaining way. For sure this is more work for the teacher but in the end the rewards of educated students are very much worth the effort. Nevertheless, a valuable lesson was learned for this teacher—make the class lectures entertaining and the students will respond better. When I began these articles for *Scary Monsters* I kept this entertainment element front and center and made the articles entertaining. And this logically starts with the article's title. In this particular case, a riff on one of my favorite films, *The Wizard of Oz*. Educate *and* entertain! I am reminded of the famous quip by the inventor, Thomas Edison, who hated exercise, who said, "the body's chief job is to carry around the brain."

"I'd while away the hours, conferrin' with the flowers ... if I only had a brain!"

Most of us instantly recognize the lyrics of that song as sung by the Scarecrow in the 1939 MGM production, *The Wizard of Oz*. By the end of that wonderful film we do know that the Scarecrow did indeed get his brain and all it took was a diploma. If it were only that easy ... just a diploma. But wait! According to some of our favorite SF films it just may *be* that easy so read on gentle readers.

12

Brains, Craniums, and Heads, Oh My!

Before we peer any closer at the portrayal of the brain in SF cinema, we need to take a little time for an introduction to the anatomy of the head. Not much, just enough to get an overall understanding of what is really there between our ears, which ideally differs from The Three Stooges scene where Moe looks into Curly's ear and sees Larry on the other side! Typically, the human brain weighs around three lbs., a nice example of good things coming in small packages, and that is a key feature of the brain: its compactness.

Our brains are composed of several major areas. The bulk of the brain is called the neocortex and what most think of as the actual brain is composed of four lobes, the frontal lobe (the front), the parietal lobe (the middle), the occipital lobe (back), and the temporal lobes (the sides). The frontal lobe is where planning, language, and speech expression occur. Taste and touch are controlled in the parietal lobe whereas vision is controlled by the occipital lobe. The temporal lobes control language and reception. Right and left handedness play a role in brain development as well as whether one is right brain dominated (artistic) or left brain dominated (analytical).

The cerebral cortex is the wrinkled outer layer of the front of the brain and its functions include the perception of sensations, learning, reasoning and memory. Another major area of the brain is the cerebellum (underneath the occipital lobe) and this coordinates the movement of muscles. The diencephalon is composed of the thalamus (the cerebral cortex relay center) and the hypothalamus (controls the autonomic nervous system and serves as an overall integration center). Buried in the cerebellum is the epithalamus, more commonly known as the pineal gland, which controls the body's various rhythms. The limbic system is composed of the olfactory (smell), amygdala (regulates heart beat), and the hippocampus (controls memories). Lastly is the brain stem and it is composed of the mid brain (controls body posture, blood pressure, body temperature, and regulates appetite), the pons (controls respiration, chewing, and taste), and the medulla. The medulla is considered the life-sustaining control center and controls the heart, respiratory, cough/gag/swallow reflexes, and digestion. Emanating from the brain and brain stem are the ten cranial nerves that coordinate the entire body. Most of these ten cranial nerves come from the brain stem, which is the main highway connecting all of the body's nerves and spinal cord with the brain.

12. Brains, Craniums, and Heads, Oh My!

Head Trauma

Trauma, either of biological or physical origin, to any anatomic region of the brain or peripheral nervous system, can result in homeostatic imbalance. When things go awry in our brains, major alterations to normal brain functions can occur. Our brains have multiple redundancies that take over when one area is no longer functioning as it should such as in head trauma. When this does occur it often takes long periods of rehabilitation for new areas of the brain to relearn simple tasks. And this should also give you a bit of an understanding that when things go wrong, or the wrong brain function takes over, then mental disorders can result, some worse than others.

Damage to the frontal lobe can cause memory defects, poor concentration, behavioral issues, depression, and other psychiatric disorders. If the parietal lobe is damaged it can result in motor issues. Occipital lobe damage may cause difficulty receiving visual input and ability to interpret visual images. Damage to the temporal lobe can lead to hearing loss, and impair visual memory. Those with cerebellum damage can suffer from improper muscle coordination, have a difficulty with balance (walking), swallowing, talking, and eye-hand coordination. Those with thalamus damage can have altered consciousness and perceptual losses and those with hypothalamus damage have hormone imbalances and the inability to control body temperatures. Epithalamus problems sometimes result in hypertension, epilepsy, and depression. Damage to the limbic system can cause loss of smell, loss of emotional control, sleep problems, loss of appetite, and memory. Those with damage to the brain stem suffer from a loss of consciousness (mid-brain damage), can be in a semi-coma (pons damage), or can be comatose with abnormal breathing (medulla damage).

Brain Surgery

Many assume brain surgery requires a higher level of intelligence, knowledge and skill than other types of surgery. Truth be told, though brain surgery is a delicate and precise procedure, it is no more difficult than most other delicate surgeries. Most brain surgeries last no longer than 6–8 hours, a timeframe similar to a heart transplant. Other surgeries, such as a hand transplant, are far more delicate and can typically take between 12–18 hours and require a team of surgeons. It is the mystique of the brain that is so compelling to make us all think so highly of this profession.

The Films

Far beyond mere brain surgery, the films in this section portray the infinitely more complex procedure of a brain transplant. All the nerves related to the senses, as well as heart and lung function, and muscle movement must be properly connected through the myriad of nerves from the spinal column to a new brain stem. Even after all that precise rewiring, if the foreign brain tissue is rejected, then the monster will not survive and it is back to the drawing board for our mad scientist.

Frankenstein (1931)

The granddaddy of all brain transplant films is, of course, *Frankenstein* (1931). During his anatomy lecture at the Goldstadt Medical College, Dr. Waldman (Edward Van Sloan) refers to two jars with formaldehyde preserved brains. Just so you know, tissues preserved in formaldehyde would not be functional and quite dead so no amount of electricity would revive them. Jar #1 bears two labels, "CEREBRUM" and, "NORMAL BRAIN." Jar #2 is labeled "DYSFUNCTIO" on one line and underneath it is the word, "CEREBRI," which is an odd label for a "dysfunctional cerebrum." During his lecture Dr. Waldman points out, "the distinct degeneration of the middle frontal lobe" on the abnormal brain that signifies a man "whose life was one of brutality, violence, and murder."

Later, when Fritz (Dwight Frye) enters the lecture room to steal a brain he initially takes a jar with the label, "NORMAL BRAIN." However, this label is clearly hand written and is therefore different (we will call this jar #3) from the one Waldman referred to as normal during his earlier lecture. After dropping the normal brain, Fritz then takes the next one that is labeled, "ABNORMAL BRAIN," and we can clearly see that this too is hand labeled making it distinct (we will call this jar #4) from that also shown earlier by Waldman which has a printed label. Each word on the handwritten label is underlined adding to its distinction. Did a lab assistant switch jars after Waldman's lecture and before Fritz entered the room? It is the brain in jar #4 that ended up in the skull of the monster.

Shortly after the monster was brought to life both Frankenstein and Waldman are having a casual conversation and Frankenstein comments that the "brain must be given time to develop." This is an interesting statement to make and shows that the good doctor does indeed know what he is talking about in that after any surgery the body needs to recover and allowing the brain time to develop (I see this as total nerve re-connection which would take time) is necessary for this. In other words, to convert dead tissue to fully living tissue via electricity and transplant will need time to develop and Frankenstein acknowledged this.

When the monster is brought out from a back room and into the lab we learn much about the monster's cognitive abilities. The Monster hears, understands, and responds to voice commands ("sit down"). When Frankenstein opens a skylight the monster looks up. Then the monster emotes with his face and hand gestures, clearly indicating that his nerve impulses are working properly.

The monster also responds vehemently when Fritz him approaches with a torch, demonstrating the rather sophisticated fight or flight response. Apparently, just a few days is enough for the monster's brain to be given time to fully develop.

12. Brains, Craniums, and Heads, Oh My!

Ghost of Frankenstein (1942)

Dr. Ludwig Frankenstein (Cedric Hardwicke) seeks to reform the monster (Lon Chaney) by transplanting a normal brain to replace the criminal brain he inadvertently inherited.

He prepares for the procedure planning to use the brain of the slain Dr. Kettering (Barton Yarborough), but unbeknownst to him, Ygor (Bela Lugosi) and Dr. Bohmer (Lionel Atwill) conspire to use Ygor's brain instead. Their plan is a brief success culminating in the marvelous moment of Lon Chaney's monster speaking in the voice of Bela Lugosi's Ygor. (In reality, of course, it is the vocal cords that determine vocal tone and quality.) Nevertheless, the effect is impressive and clearly makes a point: Ygor is now in charge. But not for long. Shortly after the brain transplant the monster goes blind, as noted by Dr. Frankenstein, due to tissue rejection. Dr. Kettering's brain was a transplant match with the monster's body whereas Ygor's brain was an obvious mismatch. In the end, it is the monster's immune response that ultimately wins out by rejecting the brain transplant.

House of Frankenstein (1944)

Dr. Gustav Niemann (Boris Karloff) is kicked out of Visaria University after trying to transplant a human brain into a dog. Seeking revenge, Dr. Niemann plans the trifecta of brain transplants by making arrangements to insert Herr Ullman's (Frank Reicher) brain into Frankenstein's monster, Herr Strauss' (Michael Mark) brain into the Wolf Man (Lon Chaney), and the monster's brain into Talbot (also Chaney).

Nieman says, "Talbot's body is the perfect home for the Monster's brain ... which I will add to and subtract from in my experiments." What is interesting is the brain surgery was done in a refrigerated cold room (with windows!) suggesting that cryosurgery is an important element of brain transplants. I wonder how many hours it took to do three brain transplants?

Human Brains into Apes

The Monster and the Girl (1940)

Scot Webster (Phillip Terry) is framed for a murder he did not commit and is executed. A scientist (George Zucco) transplants his brain into the body of a gorilla. The transplant is successful and enables Webster in gorilla form to exact revenge.

The transplantation of tissue into a different species is called xenotransplantation and the host's immune response (the ape's) would reject the transplant (the human brain). Even though apes and humans have a 98.5 percent similarity in their DNA, it is not enough to allow complete acceptance of a tissue graft. As a result, neither the ape nor the human brain would survive. Furthermore, the shape of an ape's cranium is different from that of a human so the fit would not be tight allowing the brain to slosh around causing many mental problems.

The Colossus of New York (1958)

A brain surgeon William Spenser (Otto Kruger), transplants his dead son's Jeremy Spenser (Ross Martin) brain into an unfeeling mechanical body. The resulting sensory disconnect suggests that for a human brain to properly function the nerves need to be connected to flesh and not machine. Prior to the transplantation Spenser kept Jeremy's brain on life support in a tank à la *Donovan's Brain*. The concepts of this film were quite ahead of their time, and Jeremy's robotical resurrection serves as a precursor of the *Terminator* cyborg.

Head Transplants

Head transplants have been a staple in SF films for decades. The idea itself is simple and easy for anyone to imagine, but anatomically, the head and neck are the most complicated region in the entire human body with numerous delicate muscles, blood vessels and layers of connective tissues. Any physician attempting such a procedure would immediately have his license revoked.

Frankenstein's Daughter (1958)

For an interesting twist on Frankenstein's monster, this daughter resulted from mad scientist and Frankenstein progeny Oliver Frank (Donald Murphy) transplanting the head of Suzie Lawler (portrayed by Sally Todd, 1957's Playmate of the Year) onto a stitched together male body. Transplanting female tissues into a male body is medically possible, and really not all that complicated. General tissue rejection would be the major issue to overcome. The monster does take and execute commands so he/she/it has high level cognitive abilities from the head transplant.

The Incredible Two-Headed Transplant (1971)

Doctor Roger Girard (Bruce Dern) transplants a criminal head (are there no other options?) on to the shoulder and next to the other head of Danny (John Bloom). Beyond saying that the major challenge in transplanting a head onto the shoulder of a body is making sure all the nerves are appropriately connected since there is some distance between the shoulder and the spinal column. Some nerve connections may be made but not all of them so the transplanted head would have some deficiencies. To speak the grafted head would need the lungs of Danny so there could be some

12. Brains, Craniums, and Heads, Oh My!

competition for breathing. In the film the two-headed beastie runs amok, which is enjoyable enough, with two bowls of popcorn.

Keeping a Disembodied Head Alive

So, what would it really take to keep a bodiless head alive? Before we, ahem, head into a discussion of the films, let us first take a look a relevant reality. United States patent #4,666,425 is titled, "Device for perfusing an animal head." In simple terms, this means it is a device for keeping a head alive by tube feeding. This patent describes a "cabinet," that provides both physical and biochemical support for an animal's head (read: human) that has been "discorporated" (as Mord the executioner from *Tower of London* would say, "severed from its body").

The cabinet has a collar that supports the head where tubes supply oxygenated blood and other nutrients through the neck and deoxygenated blood returns to the cabinet via additional tubes. In the cabinet the carbon dioxide and various waste products are removed from the blood, which is then re-oxygenated the blood and returned to the head along with new nutrients in a continuous cycle. This patent was filed on December 17, 1985. The critical component of a obtaining a patent through the U.S. Government is providing a working model (in legal terms this is called "reduce to practice") and demonstrating that it actually works and most likely this was an actual animal. This is intended to keep all the crackpot ideas out (such as a perpetual motion machine). So, what this all means is that keeping a head that has been separated from its body alive has been officially recognized as possible by the U.S. Patent Office.

The Films

The Man Without a Body (1957)

Wealthy Carl Brussard (George Coulouris) is diagnosed with a brain tumor so he decides he needs a new brain. He employs the services of scientist Dr. Merritt (Robert Hutton), who is known for transplanting monkey heads. In his lab we also see a disembodied monkey head being kept alive with pumps and various tubes connected to an artificial heart and artificial lung.

For his new brain, Brussard arranges for the head of the early 16th Century prognosticator Nostradamus to be removed from his grave. It is noted that the head was severed "professionally" so the larynx (voice box) and other delicate neck tissues were intact.

Nostradamus' head is revitalized through immersion in a jar of liquid with many tubes connected to life support. His head is eventually transplanted onto the body of Merritt's assistant Dr. Lew Waldenhouse (Sheldon Lawrence), who was murdered by

Surgery

Brussard. The transplant surgery is successful with the newly embodied Nostradamus easily moving around so all the nerves and tissues are appropriately connected and the head could effectively control the body.

The Head (1959)

Dr. Abel (Michel Simon) is a transplantation specialist who invents a robotic operating table (similar in principle to present day da Vinci robotic operating machines; once again, science fiction predicts real science). After Abel dies of a heart attack his assistant, Dr. Uud (Horst Frank), uses the robotic operating table and removes Abel's head, keeping it alive. The feeding tube setup seen in this film is more convincing than that shown in the similar *The Brain That Wouldn't Die*. With a nod to some sort of realism we see air being pumped through a tube to push air over Abel's larynx (vocal cords or voice box) for speech. With no air going over the larynx, as typically provided by our lungs, there can be no speech.

The Brain That Wouldn't Die (1962)

When Dr. Billy Cortner's (Jason Evers) fiancé Jan Compton (Virginia Leith; a.k.a., "Jan in a Pan") is decapitated in a car accident he connects the severed head to life support while he searches for a body to graft the head onto. Though his ultimate goal is to perform a head transplant, the plot never advances that far and all is ablaze at the end. Jan's violently decapitated head would be missing critical neck tissues such as the larynx, jugular veins, neck vertebrae, delicate muscles, and various nerve centers essential to life. Lastly, the disembodied head developed the ability to telepathically communicate with the closet monster.

Madmen of Mandoras a.k.a., *They Saved Hitler's Brain* (1963)

Adolf Hitler's (Bill Freed) disembodied head is kept alive in the South American country of Mandoras. For an interesting twist in the ever-popular keeping a disembodied head alive sub-genre, his head, complete with trademark moustache, is housed in a glass dome from which it issues a steady stream of orders. Initially, the dome is secured on some sort of control cabinet, but later it is simply lifted off the cabinet. With no visible means of life support the head continues to give commands. This immediately brings into question what the purpose of the control cabinet really was.

Donovan's Brain (1953)

The *Donovan's Brain* films deserve special mention because of the unique nature of simply keeping a brain alive with no transplant intentions involved. Our brains have two major metabolic demands of oxygen and nutrients. With the appropriate perfusion system (tube feeding) this could work but I do have my serious doubts. To monitor brain function these brains are hooked up to an EEG machine (electroencephalogram) and traditional alpha waves appear on the chart paper. It is difficult to know if these chart waves represent pain, pleasure, passive thought, or uninterested boredom.

Though there are a few versions of this film (all based on Curt Siodmak's story of the same name) this 1953 film is superior to the others. In general brain alpha waves signify a resting state with minimal activity so the brain waves are in synchronization and this can easily be measured on an EEG machine. When the brain is active beta waves result that are not in synchronization showing that mental activity is occurring. Well, the brain of Mr. Donovan is anything but resting and is quite active particularly when trying to transmit telepathic thoughts over distances made visible by the rapid movement of pens on the EEG chart.

The brain seen in this film is shown in a fish tank like container and only about half of the bran appears to be immersed in liquid. In our heads our brains are essentially encased in a fluid sack so being totally immersed in liquid is a natural state for our brains. The portion of the brain not immersed in fluid could dry out and not be totally functional. If true, then other parts of the brain could be trained to take over which may explain the telepathic abilities. One possibility is new re-trained thought pathways in the brain could result in telepathic tendencies.

The Brain (1962)

Featuring Anna Holt (Anne Heywood) and Dr. Peter Corrie (Peter van Eyck) this film's plot is essentially that of *Donovan's Brain,* in which a man's brain is kept alive in a tank, but here the portrayal of how the brain survives in a tank is more convincing. The surrounding life-support system is more authentic and this attempt at scientific accuracy adds much to the enjoyment of the movie.

The Man with Two Brains (1983)

In this entertaining comedy Dr. Hfuhruhurr (Steve Martin) falls in love with a woman's disembodied brain (voiced by Sissy Spacek) he keeps in his lab, which just goes to show that a man can love a woman for her mind.

Summary

In terms of difficulty I would rate them as follows. For surgery versus transplants the easiest is brain surgery in which the brain remains in the cranium and the surgeon does what is necessary. In the real world, this type of brain surgery is actually quite common and is mostly done on trauma and brain cancer patients. Several orders of

Surgery

magnitude more complicated would be brain transplants in which all of the various nerves need to be appropriately connected for normal function. Most complicated of them all is a head transplant. In addition to connecting the brain stem (!) there are all those tendons, muscles, larynx, blood vessels (the jugular being the most critical), and neck vertebrae bones that need to be connected too. A daunting task indeed.

So, now to the next level. What is more difficult, a brain transplant or keeping a head alive? To have a transplanted brain function normally all of the nerves as mentioned would have to be properly connected and working. A mighty tall order. For a head transplant just the spinal column needs to be attached (also a tall order), not to mention the rest of the blood vessels (like the jugular), tissues, tendons, and muscles. Of the two a head transplant would be technically easier though only our favorite mad scientists would perform such an endeavor.

Genetics and DNA

Life as we know it is all about DNA, those magical three letters that determine who you are and how your body will respond to the many rigors of life. DNA is the code of life, the blueprint, which makes each of us who we are. This DNA, or genes, is what we are born with. In the 21st century, the DNAge, we now have the ability to fundamentally alter the DNA we were born with to create new life forms and to repair mistakes or problems caused by life itself. Though some consider this a boon to mankind with the ability to significantly improve overall health there are the doom sayers who are worried that we are tampering with "things man were meant to leave alone." The ability to alter DNA is here to stay so now the question is what mankind wants to do with this knowledge. The thinking of Dr. Moreau brings all of this front and center on the stage of life.

13

The Legacy of Doctor Moreau

In H.G. Wells' 1896 novel, *The Island of Doctor Moreau*, the protagonist is a single-minded scientist intent on making the world a better place by toying with the genetic principles that separate animals from humans, creating "manimals," in his laboratory, which they dub his "House of Pain." A survivor of an accident at sea is stranded on Moreau's island and comes to believe the doctor's work is blasphemous. Eventually, the manimals mutiny and destroy Moreau.

The term, "humanized animals" or manimals was coined by Wells. As Wells portrayed him, Doctor Moreau is a very curious scientist and he indulged deeply in that curiosity. He crossed that proverbial SF line of "things man should leave alone" when he created his island of beasts. Such curiosity-driven research dominates much of SF cinema and our favorite mad and annoyed scientists are indeed obsessively curious. And if you wait long enough then science will catch up to science fiction where facts and fiction converge and blur. The age of Moreau's manimals is now upon us.

A number of movies have been based on the novel, but we will focus on three. The first and certainly the best, *Island of Lost Souls* (1933), stars Charles Laughton as a vile and cunning Dr. Moreau, whereas *The Island of Dr. Moreau* (1977) features Burt Lancaster as the good doctor. The weakest version, filmed in 1996 and also titled *The Island of Dr. Moreau*, stars Marlon Brando.

We can witness the evolution of scientific understanding in the three films. In 1933, the year *Island of Lost Souls* was released, the function of DNA was unknown. "Germ plasma," "vital humor," and "glands" were popular scientific concepts at the time of the film's production. There were ads in the pulp magazines encouraging the use of "gland extracts" as a cure-all.

By 1977, when the first *The Island of Dr. Moreau* was released, experiments with manipulating DNA and genes were underway in research labs worldwide.

By the time Brando took his turn as Dr. Moreau in 1996, high school students had been cloning DNA in classroom labs.

Further advances in genetic and stem-cell technologies could now bring such manimals imagined by Wells into being. The mixing of species is certainly possible and, as blasphemous as it may be, scientists can now create species faster than God.

There is an ongoing debate among scientists and some policy makers about the ethics of blending different species to create hybrids or what are called chimeras that

13. The Legacy of Doctor Moreau

are human-animal hybrids. Quite amazingly H.G. Wells foresaw the day when any gene or combination of genes could be inserted into an animal. An article, "Animals Containing Human Material," recently prepared by the Academy of Medical Sciences in the United Kingdom, addresses many concerns of this field of study.

Are man-beasts in the near future? And what rights should they be accorded? As the leader of the manimals, the "Sayer of the Law" (Bela Lugosi) bemoans in the 1933 film, "We are not men! We are not beasts! We are things!"

No doubt legislation will be proposed that will attempt to regulate the mixing of human DNA with animals. However, not all human DNA is the same because some genes are more important than others. But which ones? Such legislation, unless it is carefully considered, may unfortunately block important research that could impact on diseases with no known cures. It should also be pointed out that no matter the level of legislation, a single-minded scientist, on his own island, would have no legal interference to create such manimals, nor the scruples to worry about such interference.

The article by the Academy of Medical Sciences says that genetic and stem cell technologies are so advanced that the creation of manimals is certainly possible and proposes that some forms of this technology should be outright banned. This then brings up the issue of which forms should be banned. For example, a primate with a brain composed entirely of human neurons. Other potentially key traits that may be banned are reproduction issues as well as physical appearance. Furthermore, cited as being a good candidate for banning is the generation of human sperm and egg cells in an animal receptive to fertilization and bringing the resulting fetus to term, the horror of which is brought to life in the 1996 film in a graphic scene of a human-like female giving birth to ... something.

Other technology that the article suggests warrants banning is the modification of an animal brain that will lead to human cerebral function. What this means in actual practice is not yet clear. The article also cautions against the creation of a manimal that has human characteristics such as facial shape, skin texture, or speech. All of these lines are crossed by the experiments of Dr. Moreau.

Over the years in many biomedical labs throughout the world human genes have been inserted into the DNA of lab mice to study their effects. These mice are called transgenic mice or sometimes "humanized mice." Many thousands of these transgenic rodents, expressing all sorts of human genes and others grafted with human tissues, exist. These mice have been genetically engineered to specifically express human genes. One example are mice who have been inserted with all the human genes necessary to create a fully functional human immune system. This gives scientists the ability to study the human immune response without actually using humans.

This research definitely paved the way for creating new therapies and the understanding of deadly diseases. These humanized animals enable research on *in vivo* human biology that would otherwise not be possible due to ethical, logistical, or technical constraints. Such transgenic lab mice, though possessing a few human genes, nevertheless still look very much like and act like lab mice, so no real ethical boundaries have been crossed. A harmless lab mouse that has skin resembling human skin would be very useful in researching skin diseases. However, the sight of such an animal could bring about disgust and may even be frightening to others.

In addition to transgenic rodents, scientists have also created transgenic sheep and goats that manufacture human proteins in their milk. For example, human enzymes in the milk of a transgenic goat are being used to combat diarrhea-causing bacteria. Transgenic pigs containing human genes are being developed for study in transplant surgery.

In my own research I have created many rodent-human hybrids at the cellular level (it's easier than it sounds). A simple chemical, polyethylene glycol (a polymer of ethylene glycol which is car radiator fluid), is used like a glue to literally fuse rodent and human cells together thereby mixing their DNA and all cellular components. Each species, the rodent cell and the human cell, intermix and form entirely new molecules and proteins in the hybrid or chimaeric cell. It is an effective way to study the effects of genes on each other. If you will, you could call these cellular manimals.

With advances in genetics and stem-cell technology animals can now be created that resemble humans and have certain human characteristics and behaviors. Science is rapidly catching up to science fiction and presents a number of ethical concerns.

Extensive humanization of a primate brain may have certain ethical issues and a self-aware primate may have trouble expressing itself which could be agonizing for the animal. Transgenic animals with human brain material could be useful in studying neurodevelopmental disorders (like schizophrenia) by understanding how normal brain development and function occurs. And the development of embryos that mix human DNA with non-human primates will bring about other issues.

In putting human DNA into animals, transgenic mice are at one end of the spectrum, as a practice that is deemed acceptable, while at the other end are human-primate transgenic species (such as in the 1941 film, *The Monster and the Girl*, where a human brain is transplanted into the skull of a gorilla) that are obviously ethically untenable. The murkier middle part of the spectrum that generates plenty of debate.

The potential creation of hybrid embryos, relatively easy with current genetic and DNA technologies, is highly controversial, and presents a number of questions that have no easy answers. Do you bring these embryos to term or destroy them before? And when does a transgenic embryo (a manimal embryo?) cross a line and become predominantly human? And once a manimal embryo has been declared predominantly human (or even marginally human) then a whole different set of laws come into effect that are scientific, medical, ethical, social, and theological.

An example of some of the Moreau-like procedures current scientists are struggling with are the introduction of human stem cells into animals that would then integrate into the animal's body resulting in the formation of chimaeras or hybrid animals. The formation of chimaeras would take place at the embryo stage where the DNA of the two species could mix together and then randomly form some sort of new creature or hybrid. Embryo ethics is at the center of much of this debate. Are

13. The Legacy of Doctor Moreau

hybrid embryos and chimeric animals something society wants to have around? How hybrid and chimeric do they have to be in order to be considered human or quasi-human? What this all comes down to is the controlling of certain human genes (note, however, that not all genes need to be controlled) or combination of certain human genes and how they are integrated into another species. How many genes does it take to cross that proverbial line of "doing things man should leave alone"? No one really knows.

There are aesthetic considerations as well, which warrant due consideration in our image preoccupied society prone to prejudice on all matters of physical appearance. Transgenic species appear in all three Moreau films and some of the manimals are visually unappealing and provoke revulsion and disgust, like a man's face with a pig's snout or with a rat's nose.

Stem cells have the ability to change into virtually any cell in the body. This is a natural process and all animals have stem cells. Scientists are now learning how to steer the stem cells into becoming whatever they specify. For example, mouse stem cells have been converted to sperm cells, which successfully fertilized a mouse egg resulting in the delivery of normal mouse pups. Human stem cells have also been introduced into goat fetuses that produced animals with organs containing functioning human cells.

This effectively demonstrates that stem cells can be programmed to become any cell in the body and since sperm cells are the most complicated of cells this proves how sound the technology is and how promising it may be as a resolution to infertility. An infertile man could have his stem cells converted to sperm cells which could fertilize a human egg or conversely, an infertile woman could have her stem cells converted to egg cells that were receptive to fertilization.

Areas under debate by the scientific community and part of the ACHM document mentioned above have focused on three main areas of "manimal research" that should be banned outright. One is the modification of an animal brain that will lead to human cerebral function; how much cerebral function is a moving target and no one really knows for sure. The second is to establish functional human sperm and egg cells in an animal that could be fertilized. The third is to create a manimal that has human features such as facial shape, skin texture, or speech. Quite frankly, all three of these areas certainly lack scientific justification so there should be no problem in banning them. Unless, of course, you happen to be a scientist like Doctor Moreau with your own island at your disposal away from the scrutiny of all those pesky lawmakers. Many of his manimals do indeed express all the above banned features to one degree or another.

So, back to the movies, the legacy of Doctor Moreau in the 1933 film is one primarily based on the use of hormones, small biological molecules, like insulin, growth hormone, or testosterone. Doctor Moreau (Charles Laughton) performed surgery on his manimals, but the main catalyst of physical change was hormones. What is interesting about hormones is they work in a transient or temporary way and

Genetics and DNA

they must constantly be used to maintain their effects (think of a diabetic who must take the hormone, insulin, every day). Which is why Lota, Moreau's "most nearly perfect creation" reverted to her natural state when the hormone treatments stopped, Moreau futilely protesting, "stubborn beast flesh ... creeping back."

Hormones played a significant role in the 1977 film too. The changes brought about by this Dr. Moreau (Burt Lancaster) are just as transient and temporary as his predecessor. To convert the beasts to manimals Moreau needed to keep injecting hormones. Between the making of the second and third film, scientists graduated from being able to manipulate hormones, a transient control system, to exploring a more fundamental control, that which controls all life, DNA. In the 1996 film Dr. Moreau's (Marlon Brando) uses gene therapy to transform and control his manimals. At least the gene therapy procedures, if successfully carried out, would indeed result in permanent changes and not the temporary changes his hormone-using predecessors tried. And this is where the lines of science and fiction are converging in this field making the boundaries blurry instead of sharper and more defined. The world has moved from one of hormones, a transient control system, to a more fundamental control, namely that which controls all life, DNA. And when you control DNA then you effectively control life and all its forms.

Moreau describes the process of transformation as it unfolds in Braddock, the man lost at sea, who he is turning into a manimal, "Your mental processes are changing. The way you think is changing. You're beginning to think in images, concrete images. Hot, cold, light, dark, food, hunger, pain. Words becoming meaningless to you, except for the most elementary command. You've lost control. You are becoming an animal."

Moreau's manimals need to maintain homeostasis, a consistency in normal internal physiology, for both proper brain and body function. This dynamic equilibrium relies on highly responsive networks such as the nervous and immune systems and muscle coordination. In the world of manimals, some physiological traits of animals, such as fur and the ability to regulate their body temperatures (think hibernation), may be at odds as each is struggling to maintain its individual homeostasis. In this way, the two species of cells are at war with each other, and oftentimes these two are at opposite poles.

Creating transgenic animals, or manimals with quintessentially human characteristics is technically relatively straightforward, similar to using a word processing program on a computer. Think of an animal's DNA or genes as a large Word document that is broken down into chapters, subheadings, paragraphs, and sentences. Inserting DNA genes into an animal is like inserting new text into the document. You specify the location to insert the new text with the cursor and then paste it into the pre-existing text. Depending upon a variety of factors the new text could either blend in seamlessly, or awkwardly. The content could be totally unrelated to the surrounding text, or maybe even in a different language.

A similar process is involved with inserting DNA genes into animals. You select

13. The Legacy of Doctor Moreau

the area of the genome (an animal's chromosomes) you want to insert the DNA, then using a series of enzymes you literally paste the new DNA gene right into the genome text. When this new text or gene is read a variety of factors will dictate how smoothly the integration goes. The new gene (or genes) could have minimal effects or change something significant that could dramatically alter homeostasis and result in something like Moreau's manimals. Technically it is really that simple, but the ethics are anything but.

If it is determined that there is a real medical benefit then this research will move forward and society will accept it. If the risks out weigh the potential benefits then the creation of manimals might remain limited to cinematic legacy.

I suspect that over time the fine line between fact and fiction will slowly blur and future societies may come to embrace the legacy of Doctor Moreau. This then brings up an interesting image of people taking their pet manimals out for their morning walk. And after an evening out for dinner you no longer need a doggie bag for leftovers but, rather, a manimal bag!

Population Biology

14

Foods of the Gods

For the foreseeable future planet Earth is the only planet we have ready access to. Interestingly, since the creation of Earth, many billions of years ago, Mother Gaia has managed to take care of herself. From extreme events, such as mass extinctions when at least 90 percent of the known species were destroyed, the planet has rebounded and spun out new and improved life forms. Currently, on top of these life forms is mankind and if we do not take care of those below us then we are in jeopardy of losing everything. The next ten generations of humans will be a tipping point that will determine survival beyond that. Nevertheless, Mother Earth will rebound again with something else new and improved. And so goes the cycle of life.

During the past 400 generations starting about 10,000 years ago the human population has grown from an estimated five million people to seven billion. This explosive growth represents a three orders of magnitude increase. By the end of the 21st century prognosticators say the human population will be around 9–10 billion people. And since there may not be enough useable land they also say that these 9–10 billion will be the maximum this planet can sustain because of perceived food shortages. Simply put, there will not be enough food available to feed more than 10 billion people.

As the population continues to increase the availability of farmland will shrink due to urbanization and climate change, increasing the urgency for discovering new sources of nutrition and for protecting the sources we already have. The best way to protect food supplies is to ensure biodiversity, which creates an availability of a large variety of food sources. Maximizing land sharing, increased crop yields, and conservation all play a role. Agricultural communities are generally aware of this, but most of the population is not.

Many habitats are critical for the proper biodiversity of food stuffs. Natural forests, mixed woodlands, and farmlands are all necessary. Equally important are the habitats of insects and other animals that are critical to proper plant growth because it is these animals that handle such essential tasks as pollination. Many fruit growers are also bee keepers to help insure all their plants are well pollinated.

Overall, approximately 9 million types of animals, plants, protists and fungi inhabit the earth. Globally, there are 500,000 known plant species and man has been able to domesticate only around 100. Of these 100, thirty of these species provide man with 85 percent of our food and 95 percent of our protein and calories. Seventy-five percent of our food comes from only eight cereal species: corn, rice, wheat, oats, barley, sorghum, millet, and rye. Of the 4,500 species of mammals on earth, we have domesticated only 16. So, simply stated, 30 plants and 16 mammals will feed earth's many billions. Food that can grow in harsh environments will be essential as urbanization and habitat reduction of arable land increases. Plants can be genetically engineered to adapt to higher salinity (near salt marshes and mangroves), drier conditions and need fewer nutrients. Plants can also be genetically engineered for better nitrogen fixation in root nodules which would help insure adequate growth even in extreme environments.

Conservation Biology

Conservation biology is an interdisciplinary subject composed of science, economics, and natural resource management. Earth's biodiversity is studied with the aim of protecting species, their habitats, and ecosystems from erosion, collapse, and ultimately, extinction. Both benign and hostile factors are involved in proper population management. The underlining concept is to maintain biodiversity and therefore a healthy ecology. Some scientists estimate that over the next 50 years up to 50 percent of all species currently on the planet will become extinct.

Integral to proper population management are some strategic species necessary to maintain the right environmental balance. The most important species are called "keystone species" which serve as the central hub of an ecosystem, meaning their loss would cause the food chain to collapse. Next down the line are the indicator species and these are useful for observing the health of an ecosystem (think of a miner's canary used to indicate the safeness of air). Last are the umbrella or flagship species that cover multiple ecosystems and habitats and are therefore good indicators of sensitive areas.

Man's magical 30 plant and 16 animal species are all integral to population management and could be considered keystone species for man's continual survival. Even though there are plenty of backup indicator and flagship species to pick and choose from, if just one of the above species should be taken out of the food chain then much mayhem and global problems will result. Though these are serious and important issues they are, nevertheless, prime targets for some of our favorite SF films.

There is a tipping point at which, once crossed, catastrophic collapse of ecosystems will be inevitable. It is the many small changes in an ecosystem that lead to such collapse and once collapsed the recovery is painfully slow and difficult, if at all. Just before crossing this tipping point an ecosystem may become increasingly vulnerable to slight perturbations due to the loss of ecological resilience.

Population Biology

Population biology is a blending of disciplines including ecology, evolution, genetics, and statistics to provide an overview of how certain populations interact

Population Biology

between other organisms and the environment. The overall intent is to understand how various populations evolve and how they regulate their size. This information can be applied to address population concerns ranging from overcrowding to invasive species to extinction. Important data includes the growth of a given population, the dynamics and interactions of the population (competition, predation, parasitism, and mutualism), food sources to feed the population and survival strategies. Key to all this is how the population interacts with the environment. Over time populations will change and evolve to better suit their particular environment and ensure their continued growth.

To describe all of this in more detail several terms are used to naturally group individual populations. A species population refers to all individuals of a given species. Metapopulation refers to a set of spatially distinct populations among which there may be some migration. A population is a group of individuals that is demographically, genetically, and spatially distinct from other groups. An aggregation is a spatially clustered group of individuals. A deme is a group of individuals that are more genetically similar to each other than to other individuals. A local population is a group of individuals in a delimited area smaller than the geographic range of the species. A subpopulation is an arbitrary subset of individuals from within a given population. Many of our favorite SF films can be placed in these categories.

Under ideal conditions (sufficient food, water, and space) populations of any given species grow rapidly and follow a pattern known as exponential growth. Exponential growth is explosive population growth in which the total number of potentially reproducing organisms increases with each generation. However, populations of organisms will not increase in size forever. Eventually, limitations on food, water, and other resources will cause the population to stop growing. When a population arrives at the point where its size remains stable, it has reached the carrying capacity of the environment. The carrying capacity is the greatest number of individuals a given environment can sustain. Competition for resources among members of a population (intra species competition) places limits on population size. Competition for resources among members of two or more different species (inter species competition) also affects population size.

Human overpopulation is a term that means people's overall numbers exceeds the carrying capacity of their environment or habitat they live in. In simple terms this often refers to the relationship between humans and their environment (read: Earth). In some cases smaller geographical areas or even countries are referred to so the meaning is flexible. Overpopulation results from an increase in births and/or a decline in mortality rates, excessive immigration, an unstable ecology (such as those who live in extreme environments of cold or heat), and a depletion of available resources. As such, it is possible for sparsely populated areas to actually be overpopulated if the area in unable to sustain life.

Since the Black Death of the Middle Ages (around AD 1400) the population of earth has been continuously growing at an accelerated rate. The most significant increase has been in the last 50 years primarily due to medical advancements and increases in agricultural productivity. Though the population continues to grow, the rate of that growth has been declining since the 1980s. Currently, there are around 7 billion humans on the planet. Estimates place the carrying capacity of earth to be somewhere between 4 to 16 billion people, so depending on which estimates are accurate, human overpopulation may already have occurred. Somewhere between the years 2040 and 2050 the population is expected to be around 8–10 billion people.

How Does a Cell Know Its Size?

In many of the plots of our favorite SF films animals are seemingly shrunk (shrinkology) or increased in size. One interpretation of this is that our body's cells themselves are either individually shrunk or increased in size to make the animal smaller or larger. This is physically impossible. By and large all cell types more or less stick to a fairly narrow range of sizes irrespective of being either animal, plant, or other. A sperm cell is dwarfed by egg cells and both are tiny compared to some nerve cells that can be up to a meter in length but even so the range of these divergent cell types among themselves are fairly uniform and consistent. So, if we wanted to either make larger or smaller versions of the same cell type the difficulty could be quite challenging. However, if you simply wanted to make either more (to make larger animals) or less (to make smaller animals) of the same cell types, or even the entire animal for that matter, then an adjustment of hormone levels and nutrient access is all that is really necessary. Burning more calories than taken in each day will result in a smaller animal (read: weigh less). As such, this is a reduction of overall size, meaning a reduction in the total number of cells, and not a reduction of each cell itself. Cell size is a constant though the numbers of each cell type can vary considerably. Skin cells are essentially the same size though an elephant would certainly have many more than, say, a mouse.

Resources for All

No species (or organism) lives in a vacuum. Each species is dependent upon many other species for survival. Humans at the top of the food chain means that all those species below us are important for our survival. Each species contributes to the bigger picture. With so many humans, especially during the last century or two, concerns have been raised that the planet may not be able to sustain this many people in a meaningful way. So many of us naturally puts an increased demand on resources such as fresh water and food. Also as a result, many environmental problems such as global warming and pollution are increasing. This also places a strain on the environment that makes everyone suffer. There is starvation and malnutrition in some areas and the consumption of fossil fuels is increasing faster than the rate of regenerating them. What this all suggests is ultimately a decrease in living conditions and the United Nations blames much of this on the waste and over-consumption by wealthier nations.

Some of our favorite SF films have addressed the need for more efficient food production and population control, though not all of them are best suited for the future of mankind. Some of these film plots to increase food stuffs and deal with overpopulation are just downright scary. So, in the world of SF we know there are options above and beyond those found in boring reality. In some of our favorite SF films we have choices. We can make animals smaller so they eat less and therefore have more to go around. We can make food larger so there is more to feed normal sized people. We can make food more plentiful, though the same size (more efficient farming). And, according to some SF films, we can also miniaturize humans and/or increase size of foods (do both?). The best of all worlds would be perhaps to have Lilliput-sized people with Brobdinagian-sized food.

To achieve all this a number of interesting plots have populated our favorite films. There is radiation induced gigantism, hormonal induced gigantism, and genetic engineering for gigantism. Yes, quality and quantity. Increase the quality by engineering in more nutrients (such as vitamins) and increase the quantity by either making them bigger or more plentiful (or both).

The Science of Shrinkology

The Devil Doll (1936)

Paul Lavond (Lionel Barrymore), an escaped convict who was unjustly accused, seeks revenge on the three partners who swindled him and set him up for imprisonment. Lavond happens upon a scientist, Marcel (Henry B. Walthall), who has developed a method for miniaturizing animals and people as a solution to overpopulation.

Marcel describes his vision, "Millions of years ago the creatures that roamed this world were gigantic. As they multiplied, the earth could no longer produce enough food [note: not true]. Think of it, Lavond, every living creature reduced to one-sixth of its size, one-sixth its physical need. Food for six times all of us." "Lavond you know that all matter is composed of atoms and all atoms are made of electrons. Well, I found a way to reduce all atoms in a body simultaneously to any desired degree and still maintain life."

Marcel wants to shrink everything to $\frac{1}{6}$ its natural size so therefore a six foot tall man would be shrunk down to one foot high. Well, if he can go "to any desired degree," then why not $\frac{1}{12}$ or $\frac{1}{24}$? This would just make even more food available for earth's population. But what about machinery? Will that also have to be reduced?

14. Foods of the Gods

Even at ⅙ scale then how would a foot tall man operate a car or for that matter work (gigantic) farm machinery necessary to till the soil, harvest full sized corn or make canned fruit to feed all those small people? How about a ⅙ scale man handling the recoil of a full sized rifle? Also, tools will have to be reduced to ⅙ size to be useful for miniaturized people. If Marcel's invention can not do this then how will such miniature modifications to machinery and tools be done? Also, what would happen if a normal sized raindrop (or golf ball-sized hail) hit a ⅙ scale person?

To feed all these ⅙ scale people with normal sized food would require some people to stay normal sized to deal with all the logistics of growing and distributing food to everyone. And who wants to be Brobdinagian to all those Lilliputs?

To prove that his invention works, Marcel reduces some dogs to a smaller scale, and demonstrates the reduced need for food by giving just a few crumbs of bread to feed the entire pack of dogs.

Marcel and his wife, Malita (Rafaela Ottiano), have a young servant named Lachna (Grace Ford) who is mentally disabled. Marcel theorizes that by shrinking her, Lachna's mental power would become concentrated, increasing her intelligence. After his hypothesis is proven to be true, ill effects to Lachna cause Marcel to have a heart attack.

Lavond and Malita enact revenge using Marcel's invention to create a devil doll, who helps dispatch two of the three men who did Lavond wrong. The third partner confesses, thereby clearing Lavond's name.

Making Animals Larger

Tarantula (1955)

Chief scientist, Professor Gerald Deemer (Leo G. Carroll), is working on a synthetic growth hormone serum and uses it on several species, including a tarantula, to make them gigantic. In his lab out in the Arizona desert, Prof. Deemer is developing a nutrient that can help meet the food demand of overpopulation. The synthetic nutrient, "3Y" makes animals larger, but affects humans differently. As an example of how well 3Y works in animals, after a single injection a baby rat doubles in size in a few hours. A white rat was seen that had three injections of 3Y and at 12 days old was around 2.5 feet long. A guinea pig, given three injections, though at different intervals than the rat, becomes 3.5 feet long at 13 days old. A rabbit reaches full-grown maturity after 6 days. The tarantula is given six injections (where do you give an injection in a tarantula?) and on the 20th day is four feet long. During a lab accident the tarantula escapes and does not receive any more injections of 3Y, yet he continues to grow suggesting the synthetic nutrient induced permanent growth changes. It should be noted that the same synthetic nutrient, 3Y, works across various phylum, animals and insects, so there must be some common growth mechanism involved.

Population Biology

Stephanie "Steve" Clayton (Mara Corday), is doing graduate work in biology and working on her thesis, "Nutritional aspects of expanding populations," which is remarkably forward thinking for 1955. Her studies lead her to Deemer's lab to spend the summer as she says, "a lab technician, cook, student, the works."

In describing his work and the nutrient, Deemer says, it's a "completely non-organic food concentrate [perhaps the first "energy drink"?]. Medicine has lengthened the life span of people. We live longer, but the food supply remains very static. The world population is increasing at the rate of 25 million per year [actually, much more than that]. An overcrowded world. That means not enough to eat. The disease of hunger, like most diseases, well, it spreads. There are 2 billion people in the world today. In 1975 there'll be 3 billion. In the year 2000 there will be 3 billion, 625 million. The world may not be able to produce enough food to feed all these people. Now perhaps you understand what an inexpensive nutrient would mean." Dr. Hastings (John Agar) then says, "How many of us look that far into the future?" Deemer responds with, "Our business is the future. No man can do it on his own, of course. You don't pull it out of your hat like a magician's rabbit. Well, you build on what hundreds of others have done before you." Needless to say, Deemer's population predictions were quite wrong.

In further describing the nutrient, Deemer says, "It is one thing to develop a formula on paper and another to make it work. So far we've found an almost consistent instability in the material. One batch of nutrient varies sharply from the next." The FDA would find such comments very discouraging and request many more studies to remove "instability in the material." One possible clue to this instability is the use of fictitious "ammoniac" which Deemer says is a radioactive isotope that "binds and triggers" the nutrient. Not sure what this means so it could mean anything. Just so you know "binding and triggering" are common biological responses. For example, the hormone insulin binds to its receptor that subsequently triggers the use of glucose so, according to Deemer, some sort of radioactive isotope binds to a component in 3Y that triggers a biological response that causes extreme growth in the species tested.

Two people were injected with 3Y, Deemer's assistant and Deemer himself. As a result they each got the disease, acromegaly, which is a completely different response from that seen in the other species, though all are growth hormone related. It appears that 3Y enhances growth in many species, such as rodents, rabbits, and tarantulas in a uniform way but does not work in a uniform way with humans since acromegaly results.

Acromegaly is a rare disease that happens with the anterior pituitary gland produces excess growth hormone. For most people growth hormone production stops when adulthood is reached but in those with acromegaly growth hormone production continues. As a result of excessive growth hormone production severe disfigurement, often with complicating conditions, results. This disease most commonly affects adults in middle age when growth hormone production has mostly stopped. Acromegaly is a slow progression disease, often years in the making, and difficult to

14. Foods of the Gods

diagnose in the early stages since changes in external features, especially the face are slow to be noticeable. (In this respect Dr. Hastings was correct in challenging Prof. Deemer's claim that his assistant died of acromegaly in just a few days. Acromegaly takes years to develop and not a few days.) Acromegaly only affects some bones and not all so some bones grow out of proportion with others. Typically, the cheekbones expand, the forehead bulges and overlying skin is thickened (sometimes a heavy brow ridge is prominent, called frontal bossing), and the jaw is enlarged. (A good example of this in the world of SF cinema is actor Rondo "the Creeper" Hatton, who was afflicted with acromegaly in real life. The bones of his hands, jaw, brow, and hip continued to grow out of proportion to his other bones. The Rondo Awards are named after Mr. Hatton.)

For Deemer to make edible animals larger to feed more people these animals would need copious amounts of the 3Y nutrient. All of the animal treatments would involve injecting 3Y with a syringe needle so those costs and logistics must be added. As Steve says, "it's (3Y) kept animals alive who have been fed nothing else." Though Deemer called 3Y an "inexpensive nutrient" its global costs could be significant. Since it appears to give humans acromegaly then its use and distribution would have to be carefully controlled. Since it is unstable then its shelf life is unknown and samples may be active for only a short period of time limiting its usefulness and therefore increasing the costs.

At the end, the escaped tarantula is now 100 feet high and roams the Arizona desert causing the usual mayhem, until he is destroyed by Air Force napalm in a fiery finale.

Making Larger Food

The Beginning of the End (1957)

At the U.S. Department of Agriculture, Illinois Experimental Station, entomologist Ed Wainright (Peter Graves) is growing large tomatoes (about 2 feet at the radius) and strawberries (about 10 inches). Wainright is using radioactivity as a way to mutate certain plants to make them grow larger with the goal of providing food for an ever growing population. In following a lead on an incident in Ludlow, Illinois (population 150), where everyone was completely and inexplicably decimated, a reporter, Audrey Aims (Peggy Castle), asks the editor-in-chief of the National Wire Service if anyone in the area is using radioactivity. He refers her to Ed Wainright. When showing two large lead containers of radioactive samples (probably low level ^{14}C or ^{3}H) Wainwright says to Aims, "Isotopes aren't explosive." As shown, the radioactive material, properly stored in lead lined containers, appeared to be of low level quantity and not particularly dangerous.

When Wainright shows Aims the gigantic fruit, she asks, "Can you eat them?"

Population Biology

He responds, "No, not yet, but we hope to develop one day a hybrid that can be eaten." This suggests some sort of toxin or mutagen is in the plant making it unsafe for consumption. Perhaps the radioactive atoms in the plant food were incorporated into the plant tissues thereby making them unsuitable for eating. Describing his vision for the large fruit Wainright says, "This we hope is the future of the American farmer and, for that matter, all farmers everywhere."

Wainright continues with, "To most of the public these giants are freaks of nature with no practicle value." Aims then says, "How do they get so big?" Wainright responds with, "Well, radiation causes photosynthesis (no!), that is the growing process, to continue night and day. The radioactive isotopes act as sort of an artificial sun. A sun that never sets." The implication is that the action of radioactivity is such that it provides enough useable energy to continue to power plant growth even at night. Receiving sunlight energy 24/7 would not improve the efficiency of photosynthesis nor would it make plants grow larger.

When referring to his assistant Wainright says, "He's a deaf mute. Working with radiation can be dangerous. An accident last year cost him his speech and his hearing." The unstated implication is radiation was somehow responsible for his loss of senses. It is difficult to imagine what sort of radiation-induced accident would be necessary to cause someone to lose both speech and hearing. If such a dose of radiation was responsible for this then other parts of his body would have been affected too. To actually lose both speech and hearing his head and neck area would have received a significant dose of radiation (as well as his brain, eyes, blood vessels, nerves, bones, and other tissues).

When observing his assistant tilling some soil of a potted plant Wainright informs Aims, "That's a plant food of essential minerals. Keeps the plants from burning themselves up. They have to be fed constantly. Actually, the fruit would grow much larger if we didn't limit the stimulation."

Aims says to Wainright, "You're a scientist. You think in terms of cause and effect. You may see something that the sheriff missed" (in reference to the destroyed town of Ludlow). This is an interesting statement and one that is at the core of any real scientist. Cause and effect are based on known biological, chemical, and physical principles and do indeed require different thinking skills. The sheriff would use a different skill set, one seemingly based on life's experiences and more pragmatic than a scientist. A scientist and a sheriff would see things from different perspectives so both would be important.

Intrepid reporter Audrey Aims then asks Ed Wainright, an entomologist, why he is working with plants, to which he replies, "The existence and development of plants and insects are very closely related. They're highly dependent on one another. As a plain matter of fact one couldn't live without the other." All quite true. In a robust ecological environment many divergent species are interdependent. This is the core of biodiversity.

The killer locusts are "eight feet tall, some even bigger." As Wainright explains,

14. Foods of the Gods

"Some locusts must have gotten into the lab and they ate some of the plants and the radioactive plant food. Their cell division accelerated immediately. That is, they started to grow abnormally fast. They had to have a constant food supply to sustain this growth. So a couple of months ago they wandered into the grain elevator outside of town (Ludlow). When they grew to this giant size they pushed their way out."

Later, Wainright says, "the giant's wings fail to develop. They can't fly." (Earlier, Wainright said that snails, beetles, and grasshoppers, were invading them on a constant basis. Did these insects also eat the radioactive plants and plant food? If so, then did they grow to larger size?) Also, Wainright stated earlier that the plant food worked by having radiation constantly stimulate photosynthesis, which is OK for plants, but locusts do not have nor use photosynthesis (they are *not* plants!), so how this worked specifically on locusts is a mystery.

After the locusts devour everything and everyone in Ludlow, they invade Chicago looking for more food. Through the use of radio frequency to mimic the sound of the locust queen, a scientist drives the locusts into Lake Michigan where they drown.

Lower Metabolism and Increase Life Span

The Killer Shrews (1959)

On a remote island, the chief scientist, Dr. Marlowe Craigis (Baruch Lumet), is experimenting with a model intended to overcome the problems posed by overpopulation. His ultimate plan is to make people twice as small, to slow their metabolic processes and reduce their need for food thereby preventing overpopulation.

On the island he is testing his theories on the shrew (*Sorex soricidie*). Shrews are small mole-like mammals that have a long sharp snout, spike-like teeth, and voracious appetites, and unusually high metabolic rates. As captain Thorn Sherman (James Best) says, "Looks like a rat, smells like a skunk." Shrews must eat 80–90 percent of their body weight every day. Dr. Craigis exaggerates, "they must eat three times their own weight every 24 hours." Shrews have poisonous saliva and their bite can be fatal to humans. Bitten humans can die of hemotoxic syndrome from shrew saliva.

As Craigis explains, "Think what would happen if you could isolate and identify the inheritant factor in each gene." (It could mean the eradication of many diseases. Also, for you purists, since each gene is its own inheritant factor and the inheritant factor is the gene so you cannot isolate one from the other since they are both one and the same ... a rose is a rose by another name....) Craigis then adds, "Generally, among mammals, the smaller the size the higher the metabolism and the shorter the lifespan (true). I'm attempting to decrease the size by maintaining a low metabolism and result in a longer lifespan." By genetically mutating a slower metabolism Craigis reasoned (incorrectly) that the organism would therefore grow slower without being more sluggish. To prove all this Craigis is using the shrew as his model. (We never see the vivarium where all their animals are kept and maintained.) When asked for what reason Craigis responds with, "Overpopulation. Not a problem now but it will be in time. If we were half as big as we are now we could live twice as long on our natural resources." Interesting logic. If true then dwarves and other small people should be able to live twice as long as fully grown adults. There is more due to longevity than simply altered or slower metabolism. This is the stuff of science fiction. A common fallacy in SF cinema is making an extrapolation in which everything is linear. In the real world this is not the case.

A group of people are trapped on the island by a hurricane and it turns out that Dr. Craigis's experiment in shrew-shrinking wasn't entirely successful.

In the film it was mentioned there are 200 to 300 giant shrews on the island, each weighing between 50 to 100 pounds. Craigis explains, "they were the size of buckshot when they were born but their rate of growth was abnormal and they continued to grow ... they are mutants that inherited all the negative characteristics of their breed. Somehow they managed to escape...." As Craigis says, "Any unusual experiment can produce unusual results." The unusual experiment is selective mutational breeding and the unusual results were the creating of giant shrews.

Dr. Bradford Blaine, a geneticist, says "Very soon, right here in this island, there's going to be a miniature reproduction of an over populated world. And you'll see the importance of what we're working to avoid."

As the giant shrews run out of smaller animals to eat they attack the trapped humans.

As one of the very few survivors, Craigis says as they float away from the island, "In 24 hours there'll be one shrew left on the island and he will die of starvation. An excellent example of overpopulation." And a good example of survival of the fittest.

Recycling Food Stuffs

Soylent Green (1973)

This film is based on Harry Harrison's novel, *Make Room, Make Room*. The tag line of the film is, "Tuesday is Soylent Green Day." In the year 2022, the Earth is overpopulated, New York City has 40 million people, and all natural resources have been depleted. Water is rationed and fresh food is virtually non-existent. Strawberries cost $150 a jar. Food for the population is now provided by Soylent Industries.

Key elements of this film are overpopulation, the problems that causes, and ways to feed the masses. The stated overpopulation has brought about all sorts of ecological disasters. As mentioned in the film, in New York there is a "heat wave all year long" and that "greenhouse—everything burning up." Sounds like global warming. With

14. Foods of the Gods

the billions of people in the world of Soylent Green then overpopulation issues would be a day-to-day problem, if not an hour-to-hour problem. The carrying capacity of Earth has been exceeded and competition for resources is an extreme. With all these people then many would be dying on a daily basis, most likely not from old age but, rather, from some problem associated with the overpopulation issues.

William R. Simonson (Joseph Cotton) started a business freeze-drying food for commercial processing and eventually developed edible freeze-dried soylent wafers, available in three varieties: Soylent Red, Soylent Yellow, and Soylent Green. Solomon Roth (Edgar G. Robinson, in his last film role), describes the wafers as, "tasteless, odorless crud."

Soylent Red and Soylent Yellow wafers are "high energy vegetable concentrates," Soylent Yellow is made of "genuine soy bean," while the more nutritious, Soylent Green, is made from ocean vegetation (plankton). Soylent Red and Soylent Yellow wafers are "high energy vegetable concentrates." Soylent Yellow is made of "genuine soy bean." At an open market Soylent crumbs were also being sold. I always thought of their synthetic food as "CHON" (carbon, hydrogen, oxygen, and nitrogen, the main basic four atoms that make up the bulk of earth's life forms). Eventually the plankton in the oceans is depleted, collapsing an ecosystem, probably through accumulated small perturbations, so Soylent Industries decided to recycle dead humans and use this processed food to feed the overpopulation.

To process this plankton replacement, Soylent Industries takes the human remains to a waste disposal unit outside the city, away from prying eyes. When Sol Roth questions the need for the third wafer, "Why make Soylent Green?," the Exchange Leader (Celia Lovsky) responds, "It's easier. I think 'expedient' is the word." Apparently, it is easier to process human bodies than it is to repair ecosystems.

When dead bodies become scarce, Soylent Industries ramps up both their resourcefulness and their expediency. Humans who make themselves inconvenient serve as fodder for the mill—when there are riots "people scoopers" are called in to harvest bodies.

With the assistance of Roth, NYPD detective Robert Thorn (Charleton Heston) finally figures it all out: "The ocean's dying, plankton's dying. Soylent Green is made out of people. They're making food out of people. The next thing you know they'll be breeding us like cattle ... for food ... Soylent Green is people!"

Making Everything Big

Food of the Gods (1976)

This film is based on the H.G. Wells story of the same name. On a remote island in British Columbia, Mr. Skinner, a farmer (John McCliam) and his wife (Ida Lupino) discover a natural off-white creamy substance oozing out of a rock formation on their property and feed it to their chickens who grow to gigantic proportions.

Population Biology

This ooze, when mixed with some grain, and eaten by newborn animals causes them to grow to fantastic sizes. In the film four diverse species grew to a large size: two foot long wasps (insects), 18 inch long grub worms (classified as an annelid, which are organisms with a segmented body); however, though Mrs. Skinner (Ida Lupino) called them "worms" they looked more like caterpillars because of their two sets of legs, those on the front section and those on the rear section (worms do not have legs like that. Also, worms do not have frontal eyes), 6 foot tall chickens (birds; both roosters and hens), and sheep-sized rats (mammals). So the FOTG ingredient(s) were common enough to affect insects, annelids, birds, and mammals.

It should be noted that nearby plants (and probably some other insects and animals) did not grow when exposed to FOTG. This suggests that FOTG needs some sort of catalyst to activate the main ingredient(s) to make it work in creating larger organisms.

Mrs. Skinner notices that nearby plants and other insects and animals did not grow when exposed to this "food of the gods." She noted this and realized it only worked when it was mixed with chicken feed. She said it affected "only the baby chicks. Nothing seemed to affect the grown ones."

Morgan (Marjoe Gortner) then asks, "Nothing happened to the full grown chickens?" Mrs. Skinner replies, "They didn't grow none." Morgan asks, "How large do you think those rats will grow if they've gotten into that food of yours?" The answer to that question came quickly.

As noted, in addition to the chickens, worms also ate the FOTG and feed mix as well as rats and wasps. Since only young chicks who ate FOTG grew to giant size and not adult chickens this theory must also apply to the rats, worms, and wasps in which the food only affects the young and not the adult forms. This would not be a particular issue since the gestation and breeding times of these species is relatively short so there would be plenty of young ones around to eat the food.

In another example of the linear extrapolation of life seen in many SF films the giant wasps made an equally giant nest (we only saw one but there could have been more). Its weight must have put a strain on the small tether holding it to a small branch. Also, there were so many giant rats everywhere that this begs the question of what did they eat to achieve such sizes? Their voracious appetites must have been like the Killer Shrews and there must have been significant competition for food. With so many rats they probably eliminated all edible life forms on that remote island and would have destroyed the island's ecology and population dynamics.

Unscrupulous Bensington (Ralph Meeker) says to his wife, Lorna (a "lady bacteriologist"), that FOTG will be his "contribution to the world ... starving people? Going to feed them all with big chickens, the giant cows and sheep and ducks and fish and you name it." Lorna then pragmatically asks, "Sounds good but wouldn't bigger creatures have larger appetites?" Bensington replies with, "Convert it (FOTG) to plant food to feed all the big animals with big plants ... to feed all the hungry people in the world." In other words, make everything big. Well, why not?

14. Foods of the Gods

At the end of the film a couple of jars containing the mixture get washed down a small creek and into a stream where cows are drinking the water. The cows are milked and the milk ends up in small containers that makes their way into a school lunchroom where the children consume it, implying that the kids will perhaps grow into Glenn Manning–sized Amazing Colossal Children. The problem with this scenario is that the active ingredients of FOTG will first be diluted in water (perhaps significantly so), then the diluted FOTG will be consumed by cows and passed completely through their digestive systems, unchanged (!), and is subsequently excreted in their milk. This milk is then pasteurized (read "sterilized") and bottled. The assumption is during this entire process the active ingredients in FOTG remain unchanged. A mighty tall order. Maybe this process could have created a super-activated form of FOTG that could have withstood all the processing involved (like mad cow disease in which the virus makes it intact through all the meat processing steps and is subsequently consumed by man).

It is tempting to wonder what would happen if someone reduced by Marcel's invention in *The Devil Doll* ate some of Mrs. Skinner's FOTG food? Would that person then grow back to normal size? One can easily foresee a situation where people live a $\frac{1}{6}$ scale existence to save on all aspects of life but then eat some FOTG, grow to normal size, and then go on vacations, operate machinery (conscriptions like being drafted?), etc. and when done revert back to $\frac{1}{6}$ scale. An amusing scenario.

Summary

The films discussed here cover a span of 40 years, almost two generations of time. Even so, the essential elements of dealing with the human population and natural resources are common themes irrespective of when they were filmed. Our Foods of the Gods are really nothing more than a cocktail of hormones that influence growth genes; some make foods bigger, some make humans smaller, and some make humans bigger. Some of these changes are transient and some are permanent. How the hormones were overproduced, either by rays, radioactivity, or recycled nutrients just show that in some of our favorite SF films many processes can "trigger" our body's ability to produce and secrete many hormones that affect our physiology and overall growth. And some may even make species grow to fantastic proportions. Also, don't forget, if we were twice as small then maybe we could live twice as long. Or, according to Marcel, using this logic, if we were $\frac{1}{6}$ our size then we could live six times as long! And, if our food was gigantic, as made by Dr. Wainright, then our supplies would be plentiful for everyone and starvation would be history.

Radiation Biology

Radiobiology is a relatively new discipline that had its beginnings in the 20th century with the discovery of radiation. It wasn't until after World War II that the world began to understand the devastating effects of too much radiation. These real fears were effectively captured by many SF film plots of the 1950s and these fears are still with us though not as devastating as once thought. Nevertheless, the health effects of too much radiation are real and can be truly harmful if not lethal. High doses of radiation can cause dramatic burns and mutations but low doses can be useful such as in the case of radiation treatments for cancer patients.

15

Amazing Colossal Science

When I first saw *The Amazing Colossal Man* in 1957 at the Adler Theater in Marshfield, Wisconsin, it made a big impression on this five-year-old. Even then I thought it special and later in life when I understood the science I am still impressed by the amount of biology discussed in this film.

Bill Warren, in *Keep Watching the Skies*, describes it as "crass, heavy-handed and cumbersome." *Videohound* gives it two "bones" and refers to it as a "standard '50s Sci-fi film about atomic radiation." For me, I could develop a semester-long class based on the science in this film, with a syllabus encompassing everything from radiation biology, dermatology, endocrinology, cardiology, physiology, biochemical metabolism, psychiatry, stem cells and new tissue regeneration for burns.

The core of the film, written by Mark Hanna and Bert I. Gordon (a.k.a., Mr. BIG), is radiation induced gigantism. For the film's major plot, Colonel Glenn Manning (Glenn Langan) is exposed to a sudden and normally lethal dose of radiation via a plutonium bomb explosion but, instead, survives the blast and grows into a proportionally correct giant.

The story begins at the Desert Rock proving grounds in Nevada, where the military is preparing to detonate a plutonium bomb. In a nearby trench soldiers wait to receive the blast "under simulated combat conditions," but the bomb does not detonate right away. Did any of the other soldiers also receive a dose of radiation similar to Manning?

Meanwhile, a private plane crash lands near the detonation site and while attempting to save the passengers Manning is caught in the blast and takes a direct hit of plutonium radiation. He instinctively covers his eyes with his arms during the blast but the rest of his body is directly exposed to the heat and the radiation. His shirt is mostly blown/burnt off from the blast and his skin had immediate radiation burns, not to mention his lack of hair.

Radioactive Rant

The first sustaining nuclear chain reaction was performed on December 2, 1942, in an abandoned handball court at the University of Chicago's Stagg Field by Enrico

15. Amazing Colossal Science

Fermi. Less than three years later the first nuclear bombs were dropped on Hiroshima and Nagasaki which brought a rapid end to World War II. The first atomic bomb test, codenamed "Trinity," and detonated on July 16, 1945, near Alamogordo, New Mexico, had used plutonium as its fissionable material. Since this was the "first plutonium bomb" explosion then Manning must have been present at the detonation. The implosion design of "the Gadget," as the Trinity device was nicknamed, used conventional explosive lenses to compress a sphere of plutonium into a supercritical mass, which was simultaneously showered with neutrons from the "Urchin," an initiator made of polonium and beryllium. Together, these radionuclides caused a runaway chain reaction and subsequent explosion. The Gadget weighed over 4 tons, although it used just 6.2 kg of plutonium in its core. Even so, about 20 percent of the plutonium used in the Trinity weapon underwent fission, resulting in an explosion with an energy equivalent to approximately 20,000 tons of TNT, which was only one-fifth of the weapon's potential.

Fission vs. Fusion

There are two different types of nuclear reactions in which a large amount of energy is released. Nuclear fission is the splitting of a large atom into two or more smaller ones whereas nuclear fusion is the fusing of two or more smaller (or lighter) atoms into a larger one. In simple terms, fusion fuses atoms lighter than iron where fission divides atoms heavier than iron. The two major radioisotopes used for nuclear fission are uranium-235 (Ur-235) and plutonium-239 (Pu-239) and when they are done releasing their energy very stable lead-207 (Pb-207) is the end result. The ratio of radioisotopes Ur-238 to Ur-235 is used by geologists to determine the ages of minerals since their decay levels can be precisely determined. When someone says that dinosaur bones are 100 million years old they know this by the Ur-238/Ur-235 ratio.

Fission reactions do not normally occur in nature whereas fusion reactions occur primarily in stars, such as our sun, where four hydrogen atoms fuse to produce a helium atom thereby releasing a significant amount of heat. Fission reactions produce many highly radioactive particles whereas fusion reactions produce fewer radioactive particles. A nuclear bomb based on fission is also known as an atomic or atom bomb whereas a hydrogen bomb uses a fission reaction to trigger a fusion reaction, so it is much more explosive and therefore dangerous.

The atomic bomb dropped on Hiroshima on August 6, 1945, was codenamed "Little Boy" and used Ur-235 as its fissionable material and not plutonium. The second atomic bomb, codenamed "Fat Man" that was dropped on Nagasaki three days later did use Pu-239 as its fissionable material. Only after the announcement of the first atomic bombs was the existence of plutonium made public since it is a man-made radioisotope (very little exists in nature so it must be manufactured).

Radiation Biology

There are 15 different plutonium radioisotopes and collectively are a class of highly toxic and unstable chemicals whose radioactivity is measured by the number of atoms disintegrating per unit time. Plutonium radioisotopes decay to uranium by emitting many types of radiation as high-energy alpha particles (in the form of an energetic helium nucleus, He^{++}), medium-energy gamma rays, x-rays, and beta particles. The main isotope in weapon-grade plutonium is Pu-239 and it has a long half-life of 24,000 years so its potential damage is far ranging. It takes about 10 kilograms of nearly pure Pu-239 to make a plutonium bomb. Producing this requires 30 megawatt-years of reactor operation, with frequent fuel changes and reprocessing of the "hot" fuel. Therefore, weapons-grade plutonium needs to be made in special production reactors by burning natural uranium fuel.

Radiation Biology

An atom is composed of its nucleus and surrounding electrons. The bigger the atom the more components of its nucleus, mostly made up of protons and neutrons, and the more electrons. When the nucleus decays or electrons come off a lot of energy is released oftentimes in the form of radiation.

First of all, it should be noted that not all radiation is the same. Radiation is the physical process in which energetic particles or waves travel through various types of materials or space. There are two major types of radiation, ionizing and non-ionizing with ionizing being the most deadly since it has enough energy to ionize an atom; the ions produced by ionizing radiation have the potential to damage DNA and therefore can cause biological mutations many of which would be fatal. (Examples of non-ionizing radiation are radio waves, heat or visible light.) These ionized particles or waves "radiate" or travel in all directions from its source and as such are capable of being measured in three dimensions.

There are three major types of ionizing radiation, alpha, beta, and gamma and they are classified as to their degree of energy. Each type of radiation has a different biological effect because they transfer their energy to tissues in different ways. The lowest energy are alpha particles and these can be stopped by a sheet of paper. The clothes you are wearing are sufficient enough to stop alpha particle radiation. Alpha particles do not penetrate skin and therefore there is no damage to overall health. In reality, alpha particles only penetrate just a few centimeters of air. However, highly energetic alpha particles are in cosmic rays and can penetrate the body but are stopped by our atmosphere and deflected by earth's magnetic field so they are only a problem for astronauts.

Alpha radiation emitting isotopes are only a problem when ingested (breathed or swallowed) because this brings the radiation directly close enough to tissues as to be dangerous. In terms of biological effectiveness ingested alpha radiation is the most deadly because of the damage the ionizing particle leaves in its wake being so close

15. Amazing Colossal Science

to cells and tissues (and therefore, DNA). At close range alpha particles are at least 20 times more deadly at cell damage than gamma rays or x-rays. Examples of deadly alpha radiation emitters are radium, radon, and polonium.

Beta radiation particles have a higher level of ionizing radiation energy than alpha particles, though not as much as gamma particles, and can be stopped with thin metals such as an aluminum plate. Beta radiation consists of energetic electrons from radioactive isotopes and can be stopped by a few centimeters of plastic or a few millimeters of metal. Beta particle radiation occurs when a neutron decays into an atom's proton thereby releasing the radioactive beta particle. Radioisotopes that release beta radiation are sometimes used in the clinic to treat superficial tumors, so when you hear that cancer was being irradiated that means that beta radiation was used.

Another source of beta radiation is when neutrons, particles in an atom's nucleus, are released. Neutrons are the only type of ionizing radiation that can make other materials or objects radioactive by directly ionizing them. This is the primary commercial method to make radioisotopes useful in medical and industrial applications. Neutrons collide with atomic nuclei thereby creating unstable radioisotopes that release radioactivity. When a neutron knocks an atom out of a molecule the atom's electrons are left behind and the chemical bond is broken resulting in free radicals that are biologically harmful. When a neutron hits an atom's nucleus think of a billiard ball striking another ball in which all the energy of the striking ball is transferred to the hit ball causing it to move away. High energy neutrons can travel hundreds to thousands of meters through air and several meters through common solids. Nuclear reactors use water several meters thick as a shield against neutron radiation.

Gamma radiation is the most energetic, strongest, and most deadly of radiation particles and thick walls are needed to stop these particles. Gamma radiation occurs after a decaying atom's nucleus emits either alpha or beta radiation. Gamma radiation is composed of photons and these have neither mass or electric charge so they can penetrate much deeper through matter and are difficult to stop.

Another significant form of radiation released by nuclear bombs that can affect health are x-rays. X-rays are electromagnetic waves with very small wavelengths. The smaller the wavelength the higher the energy content. A packet of electromagnetic x-rays is called a photon and when an x-ray photon collides with an atom two events occur. If the photon is weak the atom could absorb the energy of the photon and boost an electron to a higher orbital level or if the photon is very energetic it could knock an electron away from its atom causing the atom to ionize. Ionized atoms are harmful to animal and human health. Generally, larger atoms are more likely to absorb x-ray photons than smaller ones. (Soft tissue in our bodies are composed of atoms smaller than the calcium atoms in our bones which is why x-rays are so effective and allow physicians to distinguish between tissue and bones when examining x-ray photos. These medical x-rays are of the low energy kind and not a real worry.)

One aspect unclear, and central to the level of radiation sustained by Manning, is how close was he to ground zero when the plutonium blast went off. If he was too close then he could have been easily disintegrated in the blast. Since he did survive then he must have been some distance away, just far enough to sustain some radiation damage but not enough to be fatal. However, it must also be pointed out that later in the film it was noted that the crashed plane and pilot that Manning was attempting to rescue could not be found, suggesting they were atomized. Manning could not have been that far away from the plane since he clearly saw it and was within running distance so there was a critical distance window that Manning was just in when the blast went off. He was not too close to get atomized, nor close enough for extensive and deep skin burns. He was at the "just right Goldilocks" distance of getting enough radiation to mutate just the right amount of DNA that altered just the right number of genes to give just the right proportionally correct growth. Perhaps a meter closer or a meter farther away from the blast could have made all the difference. Manning was just where he needed to be to get just the right amount of radiation to grow in just the right proportional way.

Health Issues

There are three principal routes by which plutonium (or any other radioisotope) can infiltrate the human body: ingestion, inhalation, or contamination of open wounds. In *The Amazing Colossal Man* Colonel Manning was not only exposed to the initial bomb blast causing severe burns, those burn were open wounds immediately subject to contamination. He also probably inhaled a significant amount of the radiation. And if the radioactive particles were in his mouth, very likely, they mixed with saliva and were subsequently swallowed, so he ingested them, too.

The stomach does not absorb plutonium very well, so ingestion is not a significant hazard because it is expelled through the gastro-intestinal system before it can do harm. The main threat to humans comes from inhalation. When inhaled, plutonium can remain in the lungs, depending upon the particle size and how easily the particular chemical form dissolves. The chemical forms that are slow to dissolve may lodge in the lungs or move out with phlegm, and either be swallowed or spit out. But quick dissolving forms would be absorbed by the lungs and passed into the bloodstream.

While it is very difficult to create airborne dispersion of a heavy metal like plutonium, certain forms, including the insoluble plutonium oxide, at a particle size less than 10 microns (0.01 mm), are a hazard. If inhaled, which Manning did, then much of the material is immediately exhaled or is expelled by mucous flow from the lung airway bronchial system into the gastro-intestinal tract, as with any particulate matter. Some however will be trapped and readily transferred, first to the blood or lymph system and later to other parts of the body, notably the liver and bones. This had to have happened in Manning's case. It is here that the deposited plutonium's alpha radiation may eventually cause cancer via the mutation of DNA. Once internalized and in the blood stream, plutonium radioisotopes can most significantly affect lung, bone, liver, and other important organs. Plutonium that reaches organs can stay in the body for decades and continue to decay and expose the surrounding tissues with harmful radiation thereby increasing the risk of cancer. Plutonium is also a toxic metal, and may cause damage to the kidneys.

People can inhale plutonium as a contaminant in dust and it can also be ingested with food or water. This is especially relevant for people who live near government weapons production or testing facilities who may have increased exposure. External exposure of plutonium to the body poses very little health risk since plutonium isotopes emit alpha radiation, and almost no beta or gamma radiation. In contrast, internal exposure to plutonium is an extremely serious health hazard. Internal exposure continues until the radioactive material is either flushed from the body by natural processes or decays. Inhaled or ingested plutonium is distributed to different organs and will remain there for days, months, or years until it decays or is excreted.

However, the hazard from weapons grade Pu-239 is similar to that from any other alpha-emitting radionuclides that might be inhaled. It is less hazardous than those that are short-lived and hence more radioactive, such as radon daughters, the decay products of radon gas, which (although in low concentrations) are naturally common and widespread in the environment.

The ionizing radiation caused by plutonium disrupts molecules in cells and deposits energy in tissues, causing damage. Both immediate and delayed health risks occur. Observable effects occurring soon after receiving very large doses include hair loss, skin burns, nausea, and gastrointestinal distress or death (this is called Acute Radiation Syndrome). Long-term risks, including increased cancer risk, are a function of the specific radioisotopes involved and depend on the route, magnitude and duration of exposure.

For those exposed to large doses of radioactivity the emergency medical care to save their lives is the first priority. Effective patient decontamination is important to limit the spread of radioactive materials in the hospital, and to prevent exposure to other patients and staff. Typically, these patients are placed in isolation wards to help minimize contamination. Treatment to reduce the internal radioactive dose consists of using chelators, chemicals that tightly bind to the radioactive metal atoms, and flush them out of the body thereby reducing the radioisotope body burden. Some forms of radiation can stay in the body for years.

To minimize their exposure to radiation, called secondary contamination, the time people spend with the patient should be limited. They should avoid direct contact, maintain some sort of distance from the source, and use shielding or respiratory protection to prevent ingesting or inhaling any radioactive contamination. Patients exposed to excessive radiation have many tests taken to monitor their health and situation. The most common test is a blood test where the number and function of the patient's blood cells are analyzed. Since blood cells readily divide, compared to other cells in the body, defects brought about by the effects of radioactivity can be easily detected. These blood tests would be done frequently.

Plutonium is far from being "the most toxic substance on Earth." Just so you gentle readers know, there are substances in daily use that, per unit of mass, have equal or greater chemical toxicity (arsenic, cyanide, even caffeine) and radiotoxicity (such as smoke detectors) than plutonium.

Dermatology or the Skinny on Skin

The study of skin is called dermatology. The primary purpose of skin is to serve as a barrier between the harsh and contaminated outside world and our sensitive and sterile internal body. The skin contains many specialized cells and structures that not only help maintain a proper body temperature, but also plays an active role in our immune system in protecting us from disease. When skin becomes severely compromised then many health issues can result. By definition, skin is the tissue above the superficial fascia where the first layers of muscle and nerves are located. Those tissues below the fascia are not considered skin. Skin is composed primarily of three layers. The upper or outer layer of skin is called the epidermis and is mostly composed of dead and dying cells and itself is composed of five different layers; as the cells move up these layers they push already formed cells further upward to the surface where they flatten and die and subsequently are sloughed off. This is a natural cyclical process and it takes about two weeks to shed off dead surface cells. The thinnest epidermis is on the eyelids (0.05 mm) and the thickest is on the palms of your hands and the soles of your feet (1.5 mm). Though Manning did cover his eyes with his forearms he did so with his palms directly facing the blast so those skin cells were hardest hit.

The middle layer of skin is called the dermis; it has the blood vessels and nerve endings. The dermis is primarily composed of 3 tissue layers and also varies in thickness—0.03 mm on the eyelid and 3.0 mm on the back. Within the dermal layer are many specialized cells such as the hair follicles, oil (sebaceous), scent (apocrine), and sweat (apocrine) glands, blood vessels and nerves. It is these nerves that transmit the sensations of pain, itch, touch, pressure, and temperature. When dermal cells die off they replace the sloughed off epidermal cells in a constant cycle.

The third lowest layer of the skin, the hypodermis or the subcutaneous tissue layer, is primarily a layer of fat and connective tissue. The larger blood vessels and nerves are located here. The size and thickness of this layer varies considerably throughout the body and from person to person. For Manning, it is unknown how much of this third layer was destroyed by the plutonium blast. And, of course, if any of these three major layers of skin were destroyed then complicated health issues can result and recuperation may take some time, perhaps several months for healing.

For burn victims, including radiation burn victims, there are two main options, depending upon how much skin can be salvaged from what was left behind. Either a skin graft can be placed over already existing skin, or over the fascia directly. Based on the image of Manning just after the plutonium blast, quite a lot of his skin was

15. Amazing Colossal Science

still present since we could see very little of the underlying fascia muscle layer. As such, this could serve as a good foundation upon to which to place the necessary skin grafts. Even so, it is unclear which layers of Manning's skin were damaged and which layers of skin needed replacing. As a doctor says, "third degree burns on almost 100% of his body surface and the man still lives."

Since Manning took a direct radiation blast to the front of his body, this would suffer the most radiation-induced burns. The skin on the back of his body would not have had such severe burns and could have potentially served as skin grafts for the rest of his body. Even so, the surgical procedure to begin to restore Manning's destroyed skin would have taken many long hours.

Since his skin healed so rapidly it appears that only the topmost superficial layer of skin, the epidermis, was burnt off by the heat of radiation and not the underlying dermal layers, leaving behind enough undamaged tissue to serve as the basis for self-repair.

Body Fluids

A doctor examining Manning says, "he's already lost enough body fluid to be fatal." Skin keeps fluids within the body and when skin is destroyed or becomes leaky then important and vital body fluids can ooze and leak out and can be life-threatening if severe enough. Body fluid or bodily fluids are liquids such as blood, saliva, sweat, urine, bile, milk, etc., that are located inside bodies. The dominant body fluid is water and approximately 60–65 percent of water in our body is contained within individual cells whereas the remaining 35–40 percent of body water is outside cells. There are three main compartments of fluids within our bodies. One is the plasma of circulating blood (extracellular fluid), the second is the interstitial fluids between cells, and the third is the fluid located within cells (intercellular fluid). When the skin is destroyed as seen with burn victims then all of these compartments are compromised and lift-threatening fluid loss can be an issue.

Radiation Induced Shock

In the movie, one of the attending physicians comments that Manning will most likely "die of shock before morning." Shock is a life-threatening condition that occurs when the body is not getting enough blood flow and this can damage multiple organs. People in shock usually have extremely low blood pressure. Shock requires immediate medical treatment and can get worse very rapidly. Shock can be caused by any condition that reduces blood flow, including heart malfunctions (like a heart attack or heart failure; cardiogenic shock), low blood volume (as with heavy bleeding or dehydration as seen with burn victims), physical changes in blood vessels (as in infections, such as septic shock or severe allergies; anaphylactic shock), some medicines, spinal injuries (neurogenic shock), and heavy external or internal bleeding (hypovolemic shock). Another example is toxic shock syndrome from an infection. Depending on the specific cause and type of shock, symptoms will include one or more of the following: anxiety, bluish lips and fingernails, chest pain, confusion, dizziness, pale and clammy skin, low to no urine output, profuse sweating, a rapid but weak pulse, shallow breathing, and eventually unconsciousness. Throughout the course of the film, Manning experiences just about every one of these symptoms.

Medical Care

From the moment Manning is brought into a hospital after the blast he undergoes various medical tests. He is a thoroughly studied and well cared for patient. A few times it was noted that he had a rapid respiration (a symptom of his medications to counter the shock) even though the room was kept at a balmy 82°F.

The placing of moist bandages over Colonel Manning's burnt skin accurately represents standard care for burn victims. This is to primarily prevent further fluid loss and dehydration and to lessen the likelihood of infection. To address infections and fluid loss Manning was given "penicillin and cortisone around the clock." At one time, especially during the 1950s era when *The Amazing Colossal Man* was produced, penicillin was considered a wonder drug. Now, several decades later here in the 21st century, penicillin is more cautiously prescribed due to overuse resulting in the development of resistant strains (such as MRSA, pronounced "mersa," which is *m*ethicillin *r*esistant *Staphylococcus a*ureus bacteria; methicillin is an analog of penicillin) making it now a "no wonder drug." Cortisone, a hormone, suppresses the immune system thereby reducing inflammation and attendant pain and swelling at the site of the injury, but it also can have serious side effects such as diabetes and bone problems.

Right after Manning was bandaged he was taken into an isolation ward and kept under an oxygen tent, another common practice for burn patients. The sterile environment helps prevent further infection, and protects the nurses and doctors from being contaminated. Next to his bed hangs a bottle of blood to replenish what he lost. The next morning when a nurse looks in on the patient she notices his remarkable skin improvement. The hanging bottle of blood is nearly empty. Is this the same bottle? Manning should have received bottle after bottle throughout the night. Both units of blood, and blood plus fluids such as plasma and a glucose solution to help provide nutrients and replace missing fluids.

The nurse, when questioned, said Manning had "cortisone administered continuous through the night in plasma containers." This means Manning received the cortisone treatment mixed with plasma fluids via intravenous (IV) tubing. Plasma contains many vital nutrients, electrolytes, and proteins to help stabilize the circulatory system and help prevent shock.

15. Amazing Colossal Science

After noticing Manning's remarkable improvement the physicians removed the bandages from his head and upper to discover, "he developed new skin," without "even a scar." A sheen of sweat and natural body oils on Manning's face and upper chest, indicate his new skin is still undergoing development.

The Heart of the Matter

The human heart is located in the middle of your chest between the lungs and behind and slightly to the left of your sternum or breastbone. The heart is encased in a double-layered sac called the pericardium that not only allows attachment via ligaments to the spinal column for support but also provides the freedom for the heart to move as it beats.

The heart of a normal sized man is about the size of your fist, typically between 7 to 15 ounces (200 to 425 grams) or, as Dr. Paul Linstrom (William Hudson) would say, is about the "size of the distance from his nose to his chin." In each day the heart averages 100,000 beats and pumps about 2,000 gallons (~7,570 liters) of blood. At the end of a normal life span the heart probably beats more than 3.5 billion times.

Human hearts have four chambers. The upper chambers are called the left and right atria and the lower chambers are called the left and right ventricles and all four are separated by a muscle called the septum. The left ventricle is the largest and strongest chamber because it is this chamber that pushes blood out the heart and into the rest of your body. And that is the main job of the heart, to pump blood throughout the body to the organs, tissues, and every cell in your body. (Blood delivers oxygen and nutrients to every cell in your body and removes waste products including carbon dioxide from those cells.) By the force of the heart beating blood is pumped through a complex network of arteries, arterioles, and capillaries and subsequently returned through the vein and venules network. If all the arteries and veins in your body were laid out end-to-end they would be around 60,000 miles long (96,500 kilometers). This is long enough to circle the earth more than twice. And for the 60 foot tall Manning this network of blood vessels would be even longer! This means that Manning's small heart would have to pump extra hard to adequately deliver all the necessary blood to those many miles of his circulatory system.

The Circulatory System in Relation to Growth

Dr. Linstrom tries to further explain Manning's condition to Carol, Manning's fiance, by saying, "all the parts are enlarging at the same ratio, except for the heart. When it was normal size the heart measured approximately the size of the distance from his nose to his chin. At the size he is now the heart measures the same at the distance from his lips to his chin. In other words, the heart has increased one half as

Radiation Biology

much as the other parts of the body. All the parts of the body consisted of millions of tiny cells that were rapidly and uncontrollably multiplying. Today, we learned this theory does not apply to Glenn's heart. Its growing but at a much slower rate. The reason for this is technical [note: never trust a scientist when you hear this. It is his job to explain the material to even the most simplist of people]. To give you a simplified layman's explanation ... since the heart is made up of a signal cell [note: in the above mentioned book, *KWTS*, author Bill Warren appropriately called this comment, "hogwash"], for all practical purposes [for NO practical purposes!], instead of millions of cells like the rest of the organs of the body, it's reacting in an entirely different manner to this unknown force or stimulus that's behind this whole thing." Quite a mouthful.

After listening to Dr. Linstrom's explanation, Carol responds, "No wonder he's always complaining about these sharp pains in his chest." The doctor continues with a grim prognosis, "His heart won't be able to carry the load any longer. Then he will die ... his mind will go first and then his heart will literally explode," which makes Manning's complaint, "the beating of my heart gets louder and louder," all the more ominous.

In Manning's case, with a small heart, in addition to the rest of his body's needs, not enough blood was being pumped to the brain so the colossal man also suffered brain trauma and became psychotic as a result. The brain needs a constant supply of nutrients (mostly in the form of glucose as a metabolic energy source) and oxygen to function normally and if deprived of them for more than eight minutes this can then result in permanent brain damage. In Manning's case he had a slow reduction of brain oxygen as his body grew but proportionately not his heart so he would have had slow episodes of brain injury that only increased in severity over time as he grew.

Food and Water

At the hospital Colonel Manning was fed intravenously. From the hospital Manning was transferred to the U.S. Army Rehabilitation and Research Center in Summit Nevada. Manning is "growing from 8 to 10 feet a day. Eighteen feet tall, tomorrow 26 feet, 35 feet maybe 40," so solid food, and lots of it, is now essential to his survival. A civilian truck arrives at the research center with "25 sides of beef," which would be a swift uptick to supply much needed calories. Manning is also shown drinking from a large container of water. Initially, his urine output would have been diminished due to all the cortisone he was getting to help alleviate his condition. After Manning was taken from the hospital to the isolated government facility he was no longer seen with an IV bottle so he would have been off the cortisone treatments by then. Later, we see Manning eating a large cooked turkey so he was consuming larger than average portions, but to maintain an eight to ten feet a day growth rate, he would need to eat and drink continuously. As a result of his excessive growth Manning was bald with no visible body hair except for his eyebrows and eyelashes, a condition called alopecia.

15. Amazing Colossal Science

Since Manning had no body hair that would suggest that not all of his cells are behaving normally so the radiation did cause some permanent damage to some cells.

In explaining Manning's condition to Carol the doctor says, "the body is like a factory, continually producing new cells to replace the older cells, damaged cells, or destroyed cells. Now this happens in all the different parts of the body. Bone cells grow new bone cells, skin cells grow new skin cells and so forth throughout the body. New cells replacing the damaged ones ... it is this delicately balanced process of new cells replacing dying cells or damaged cells that is causing the growth problem with Glen." Carol wants to know, "How can this make his whole body grow?" A fair question. The doctor replies, "this process is out of balance. For some unknown reason new cells are growing at an accelerated or speeded up rate while at the same time the old cells are refusing to die. This is what makes Glen grow. That's what made the new skin." This is essentially the same principle seen in cancer in which new cell growth isn't balanced by other cells dying off. Eventually the build up of too many of one cell type can cause death. In cancer, one type of cell grows unchecked, but in Manning's case all of his cells are growing faster and none are dying off. To underscore this moment in the film it was noted that Manning was now over 30 feet tall and weighed 2,987 pounds.

Since Manning's heart did not grow proportionally with the rest of his body it was under a lot of stress to pump all that blood through his body. This explains all the chest pains Manning experienced. However, since this is a SF film, maybe all his rearranged and mutated DNA from the blast could have provided him with another survival advantage. So you know, python snakes have the ability to increase their heart size after a meal by placing more fat or lipid into specialized heart cells. Pythons with a bigger heart can then "feed" the rest of their body and more efficiently metabolize their newly ingested meal. After the feeding/digesting period then the python's heart reduces back to a smaller size. In Manning's case, with just the right number of mutated heart genes, his heart could also have increased after a meal and once digested then reverted back to a smaller size. When the doctors examined Manning perhaps they tested him just before big meals (such as "25 sides of beef") when his heart was smaller and not yet increased due to the lipid metabolism after a large meal. In SF films mutated DNA can do anything.

Galvanic Skin Response

Galvanic reflexes performed by Dr. Coulter were said to be "inconclusive." This is not a too surprising result. The galvanic skin response (GSR) is a method purported to measure three physiological reactions. The first is muscular activity; GSR can measure the bioelectrical changes in muscle. The second is vascular changes in which the dilation or constriction of blood vessels is measured. The third is secretory changes (sweat gland activity) of measuring the electrical conductance of the skin in response to reactions to stimuli, such as anxiety or stress. And the electrical conductance of the skin varies with the moisture level of the skin so those with higher skin moisture (more stress and anxiety) have a measurably different GSR than those with minimal skin moisture (less stress and anxiety). The skin's sweat glands control the skin's moisture level which in turn is controlled by the sympathetic nervous system so a galvanic response of the skin is an indication of physical or psychological arousal. To measure a GSR, or a change in the electrical properties of skin, either the electrical resistance of the skin can be measured or the weak currents generated by the body can be measured. Also, the GSR can slow with age. Currently, the most common use of a GSR is from polygraph tests, also known as a lie detector, which can record bodily changes associated with stress (read: lying).

In Manning's case the GSR was probably used to measure how he responded to the various medicines the doctors used. All in all a poor indicator but, nevertheless, the one used in this film. If the GSR was high, indicating a higher skin resistance or excessive moisture, then Manning was possibly responding to the treatment. If Manning's GSR was low then this would suggest he was not responding. Since Manning's GSR results were "inconclusive" then it is difficult to say what his skin moisture content really meant or what it indicated. At various times during the film we see Manning with excessive sweat on his face so his skin moisture level must have been high. With his rapid growth of "8 to 10 feet per day" then his metabolism was working overtime and highly metabolic states are stressful on the body and do result in skin moisture buildup. Even so, all of this was seen as inconclusive.

Sulfa Compounds

Here is an interesting verbal exchange between Dr. Linstrom and his colleague Dr. Coulter. After a long night of research Dr. Coulter says, "the answer is in the bone marrow. We were so close we couldn't see it." Dr. Linstrom immediately sensed what he was saying and responded with, "inject sulfa hydral compounds into the bone marrow." [Note his use of the plural of that word.] To keep all you gentle readers updated, sulfa drugs were first discovered in 1935 and consist of a family of antibiotics that are used to treat bacterial and some fungal infections. Sulfa drugs work by interfering with bacteria and fungi metabolism and were considered "wonder drugs" before penicillin was developed. Sulfa drugs concentrate in the urine before being excreted which is why they are primarily used today to treat urinary tract infections. Lastly, though there are a number of sulfa drugs available none of them are named "sulfa hydral" so it is anyone's guess as to what that is and what it does to bone marrow.

Pituitary Gland

In humans, the pituitary gland or hypophysis is a hormone gland about the size of a pea, though located at the bottom of the hypothalamus in the lower center of

15. Amazing Colossal Science

the brain, it is not a part of the brain. The pituitary gland consists of two components, the anterior pituitary and the posterior pituitary, and in total secretes nine different hormones. The hypothalamus releases various growth factors to the pituitary gland which in turn stimulates the release of the pituitary hormones. As such, the pituitary is known as the "master endocrine gland." Some of the hormones the pituitary gland secretes are the somatotrophins (also known as HGH or human growth hormone; more on this below), thyroid-stimulating hormone (TSH), endorphins, prolactin, melanocyte-stimulating hormone (influences skin pigmentation), vasopressin, and oxytocin (increases labor contractions in women).

Using high frequency to stimulate the pituitary gland will in reality serve no purpose. However, Dr. Coulter reasoned that by stimulating the pituitary gland this would in turn reverse the effects of the growth hormone thereby causing a decrease in an animal's size. Coultler reasoned that if HGH can increase a human's growth then stopping the production of HGH will stop Manning's growth and could even reverse the growth (not true).

The Science of Shrinkology

His proof of this was his shrinkology experiments by making smaller camel and elephant animals in testing his hypothesis. Shrinkology is the study of being and becoming small. Film examples of this are, *The Devil Doll* (1936) and *The Incredible Shrinking Man* (1957) in which full sized animals, including humans, are made smaller or shrunk. In going small there are two ways to go. If the number of cells in an organism are the same when it shrinks then its weight will be unchanged; the cells are simply just smaller, more compact and therefore more dense (think of the DC comics character, The Atom, who can reduce his size but not his mass or weight). However, shrinking by having fewer cells of the organism, such as if someone reverted back to adolescence when smaller in stature, then he will weigh proportionately less. Without trying to invoke some sort of weird biology or physics the second version of fewer cells would be preferred because cells remain cells and not some sort of super dense version.

To test his hypothesis of injecting sulfa hydral compounds directly into bone marrow (to stop growth) with subsequent high frequency treatment of the pituitary gland (to reverse growth) Dr. Coulter experimented on larger animals. He made smaller versions of a camel (a two humped camel as seen in the film is called a Bactrian; one humped camels are Dromedaries) and an Indian elephant (their ears are smaller than the African elephants) by reversing growth using his high frequency treatment. For this to actually happen there would have to be a unilaterally uniform decrease in size of all parts of each animal.

This then brings up the question that with small brains do they have the same mental capacity as full grown adults? This pertains to the above shrinkology discussion of whether smaller is a qualitative or a quantitative difference: smaller cells or fewer cells? It would be biologically easier to create larger versions of a growing animal (keep giving them growth hormones) than it would be to make a fully grown animal proportionately smaller in size. After all, this is what is done in the animal husbandry industry where many species, such as cows and steer, are made larger (and not smaller) through the use of growth hormones.

The work of doctors Coulter and Linstrom at the research center was done in a well-equipped lab. Present were copious amounts of glassware, several lit Bunsen burners, two (average) microscopes, and several cages of rabbits. A back room houses a camel and an elephant. Much of the glassware was for distillation setups indicating the good doctors were analyzing and testing small chemical compounds, such as "sulfa hydrals." Later it was shown that Coulter himself had a separate lab where some sophisticated electronic pieces of equipment were kept as well as a bench well cluttered with the work at hand. The electronic equipment could have been used in generating the necessary high frequency to alter the pituitary gland.

While Manning was ravaging Las Vegas a colonel says, "10 times the height of a normal man, weighs about 18,000 pounds." Ten times the normal height would put Manning around 60 feet tall weighing nine tons. That is quite a load on those ankles. And quite a jump from an earlier 30 feet tall and 2,987 pounds so this represents a doubling of his height and a six-fold increase in his weight. It would take a lot of "25 sides of beef" to do that.

Give Him the Needle

The large hypodermic needle is in need of discussion. To inject the sulfa hydral compounds directly into Manning's bone marrow they would need a long needle to penetrate the few feet of skin, tissue, muscle, and bone in his lower leg. It is amusing that what was used is simply a large version of a standard syringe.

However, it is noted that for fluid volumes on the syringe cylinder the words, "3 quarts" is the maximum indicated volume. Quaint. Hypodermic needles are labeled with volumes using metric nomenclature such as milliliters or liters and not the antiquated avoirdupois system of pints and quarts. This prop was made by the amazingly creative and ever resourceful Paul Blaisdell.

Mental Decline

During Manning's growth several progressive mental changes occurred and during the course of the film his mental decline becomes increasingly apparent and dangerous. His increased size with a suboptimal heart meant not enough blood and therefore not enough oxygen was getting to his brain. Over time a slow decrease in

15. Amazing Colossal Science

brain oxygen could give rise to slowly developing mental issues. His self pity and other behavior changes, such as his dismissal of his fiancé, would also be in keeping with his coming to grips with his situation. Ultimately, Manning goes on a psychotic rampage in the city of Las Vegas indicating his mental deterioration had significantly changed his behavior. During this he was seen grasping his head and wincing, suggesting headaches and an inability to cope with his situation so he released his frustrations on the town and people that ultimately resulted in his being "destroyed" at Boulder Dam (its original name was later changed to the current Hoover Dam). Note: if Manning's body was still radioactive and contained radioactivity (highly likely), then some of this from his body would have contaminated the Colorado river and everything else downstream for centuries.

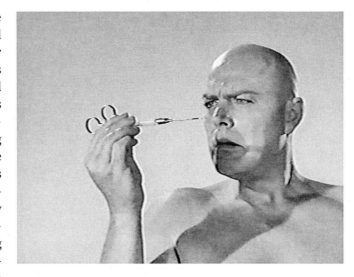

In *The Amazing Colossal Man*, the title character (Glenn Langan) examines the Paul Blaisdell–made syringe after being injected with the drug sulfa hydral.

Summary

As a result of a good Samaritan act Colonel Glenn Manning was directly exposed to a plutonium bomb blast and was literally showered with radioactive alpha, beta, and gamma particles as well as x-rays and other radioisotopes. These radioactive particles were inhaled, ingested, and entered open burn wounds and caused Manning's DNA to sufficiently mutate thereby making him grow into a proportionately equal giant, eventually becoming 60 feet tall. Unfortunately, Manning's heart was growing at a slower pace and could not keep up with the demands of his gigantic body. As a result of decreased blood flow from the smaller heart Manning suffered many physiological and psychological traumas that ultimately resulted in his being shot off Boulder Dam and into the Colorado River (and he subsequently surfaced in the sequel).

THE NURSE IN SCIENCE FICTION FILMS

This chapter was written in collaboration with my wife, Donna Glassy, B.S.N, R.N., C.M.S.R.N., a first class nurse. Donna has been a registered nurse for over 44 years and we've been married over 43. While watching an uncountable number of SF films over the years with me, Donna has kept pointing out the innumerable scenes with nurses, which otherwise I probably wouldn't have consciously registered. This helps to underscore how seemingly "invisible" SF nurses are. We take them so much for granted that they almost seem to disappear.

There are very few professions where the person is instantly recognized by what they wear. Seeing someone in a nurse uniform immediately brings much gravitas and a large background of implied information, skills, and talent. For ease of discussion when we say nurse we are referring to a registered nurse or "R.N."

No doubt the nurse is the true hero of the hospital. Nurses are often the first people patients interact with when they enter care and the last when they leave. Nurses aim to calm our nerves, soothe our wounds, provide crucial care, and serve as patient advocates.

The nursing profession is a part of the health care sector that focuses on the individual, their families, and their communities so they can obtain, maintain, and recover their own quality of life and optimal health. Nurses' approach to patient care, their training, and scope of practice separates them from other health care providers.

Though many nurses provide care within the overall scope of physicians' practices they are also permitted to practice independently in a variety of settings. Historically, the public image of the nurse as a care provider has been one under the direct direction of an attending physician. In the post–World War II era the nursing profession has diversified to incorporate advanced and specialized credentials, such as clinical nurse specialists and nurse practitioners, so the traditional role of provider is changing.

The nursing profession involves a wide variety of applications and work. To better understand what nurses do we can break down their responsibilities and duties to a few categories. Nurses help develop a plan of care with the patient, they work collaboratively with physicians, therapists, the patient and the patient's family including others, such as dieticians, focusing on treating the patient's illness (or maintain

The Nurse in Science Fiction Films

wellness) and improve the overall quality of life. Nurses like to tell patients that their body is their best medicine. A wellness of both body and mind for their patients is the goal of nurses. With advanced practice training the nurse can diagnose health problems and prescribe some medications and other therapies, depending upon individual state regulations. As a part of an interdisciplinary health care team the nurse professional helps coordinate the patient care both interdependently and independently.

16

History of Nursing

Before the establishment of modern nursing, nuns and members of the military served as care givers to the sick and wounded. Ancient history has many examples where others helped those in need and provided succor. This tradition was carried out as civilization advanced into Medieval Europe, the Middle East, and the Far East. However, the current standard of care expected of the nursing profession has only been around for the past 150 years in the U.S.

Florence Nightingale

Considered the founder of modern nursing, English born and raised Florence Nightingale (1820–1910) served as a manager of nurses to help wounded soldiers during the Crimean War (1853–1856). As a result she helped propel nursing into a bonafide profession. In 1860 she established a nursing school at St. Thomas Hospital in London which was the first nursing school in the world. Newly minted doctors take the Hippocratic oath whereas newly minted nurses take the Nightingale Pledge, named in her honor. During the late 19th Century she became an icon of Victorian culture and was known as "The Lady with the Lamp" in reference to her making rounds to see wounded soldiers at night.

Other Pioneering and Influential Nurses

Lillian Wald (1867–1940) taught a class about at-home nursing and good hygiene to immigrant women on Manhattan's Lower East Side in the 1890s. After witnessing first-hand the unsanitary conditions and lack of adequate medical care in the tenement neighborhood, Wald was moved to found the Visiting Nurse Service. Two years later in 1895, with government and financial support, she moved the VNS of New York into a larger building, founding the Henry Street Settlement House—a community center offering comprehensive assistance services. As a result, Wald and her nursing staff became the first public health nurses in the U.S. Wald was instrumental in getting nurses placed in American pubic schools and also helped establish the National Organization of Public Health Nursing, the National Women's Trade

Union League to advocate for working women, and the Children's Bureau to end child labor.

Nurse Clarissa (Clara) Barton (1821–1912) founded the American Red Cross and in any international incident where aid is necessary it is the Red Cross who is always there first.

Martha Jane ("Calamity Jane") Cannary (1852–1903) is now primarily known as a member of Buffalo Bill's Wild West Show where she appeared in her buckskins and recited her adventures but she was also a nurse. She is said to have always exhibited kindness and compassion to others, especially to the sick and needy.

Other pioneering nurses include Mary Breckinridge, Mary Ezra Mahoney, Walt Whitman (the poet), Florence Guinness Blake, Margaret Sanger and Dorothea Dix.

Types of Nurses

As the nursing profession has evolved through the decades the responsibilities and duties of the nurse has expanded. So much so that most nurses in the 21st century have a speciality. Examples are home nurses, surgical nurses, hospice nurse, labor and delivery nurse, administration, clinical nurse manager, nurse anesthetist, critical care nurse, patient educator, nurse practitioner, licensed practical nurse, and school nurses. Though male nurses are in all branches, in SF films, they tend to mostly be in the mental health services. No one type is necessarily better than another since all deal with improving both body and mind health.

Nurse Education

When Florence Nightingale founded the first school of nursing most of the education curriculum consisted of practical issues of the best ways to take care of and manage patient health. Emphasis was on hygiene and the completion of tasks. Initially, nursing was seen as an apprenticeship often undertaken in religious institutes such as convents to educate young women. Nurse education consists of integrating a broader awareness of other medical disciplines emphasizing inter-professional education to make better clinical and managerial decisions. Orthodox training does emphasize more practical skills but also emphasizes the "handmaiden" relationship nurses have with physicians. From this has evolved the current strategies of health care workers as equals who contribute to a team effort in the care and management of patients, both those sick and well. The emphasis of equal health care managers gives nurses the confidence needed to face any crisis.

Nurse education involves both the theoretical as well as practical training with the purpose of preparing them for a life as professional. This education is provided

16. History of Nursing

by experienced nurses and other medical professionals qualified for educational tasks. Initial education emphasizes general nursing and then gets into more specialized areas such as pediatric, post-operative, or mental health nursing. There are even journals on nurse education such as, *Nursing Education Perspectives* and *Nurse Education in Practice*.

Though many nurses obtain their degrees through typical four-year university level education it should be noted there are multiple entry levels into nursing. And this has led to some level of confusion with the general public, not to mention other health care professionals. Historically, the earliest schools of nursing offered a diploma in nursing and not an actual academic degree. So far as it is known Yale School of Nursing became the first autonomous school of nursing in the United States in 1923. In Europe in 1972 the University of Edinburgh was the first European institution to offer a nursing degree. The earliest schools gave diplomas in nursing and not actual academic degrees. Community colleges began to offer associate of science in nursing degrees and eventually universities picked up the pace and began to offer bachelor of science in nursing degrees. (It should be noted that Donna Glassy has a B.S.N. degree [from the University of San Francisco] as well as an R.N.). Soon to follow were programs that offered Master of Science as well as Doctor of Nursing degrees at the university level. Currently, there are even online programs for advanced nursing degrees. For nurse educators there must be a balance of practical preparations and the need to manage healthcare and the best way to do this is to have a broader view of the nurse practice. In this respect the nurse is a lifelong learner who can readily adapt efficiently to new changes both in the practice and theory of nursing.

Adapting to Changing Technology

The medical field is constantly changing so all involved in the health profession must readily adapt to new advances. As technology evolves and requires new procedures, the nurse was there to make sure the patient as a human being was taken care of. In this respect, the nurse has kept up with the science in her day and made sure the new developments helped the nursing of patients.

Patient's rooms have changed with new technology and the nurse adapted. Improvements in instruments and computers have all been readily incorporated into nurse care. Even so, it is sometimes the simple act of hand-holding or being present that provides more comfort and quality of life than any machine can give.

The introduction of computers into the nursing profession started during the early 1970s. Donna first used a computer in 1973, right out of nursing school. It subsequently took about 20 more years for computers to become a dominant element of nursing care and patient tracking. And here in the 21st century hospital the computer is almost as important as the patient.

The Nurse in Science Fiction Films

Nurse Responsibilities

Over the many decades the responsibilities of the nurse have significantly changed. At first, the responsibilities and duties of the nurse were simple and reflected what the attending doctor ordered. The nurse has kept pace with new technology and procedures and here in the 21st century she does essentially what doctors used to do not that long ago. Also, the duties of a nurse vary according to her area of expertise. The profession of a registered nurse comes along with complexities and major responsibilities. The nursing profession is a challenging vocation that demands versatility and alertness.

The primary duties of the nurse include assessing a patient's health problems and needs, conducting physical exams, encouraging the wellness of patients by promoting a wide range of services, observing and recording patient's behavior, coordinating with physicians and other healthcare professionals to create and evaluate customized care plans, caring for and educating patients and their family members, assisting in the diagnosis of disease through analyzing the patient's symptoms, maintaining proper reports on patients and their medical histories, dispensing medications, maintaining a safe and hygienic work environment, providing assistant care during medical emergencies (like car accidents, burns, and heart attacks), providing guidance in health maintenance and disease prevention, preparing patients for examinations, resolving patient's problems, assisting doctors and carrying out their orders, and recommending various drugs and forms of treatment for optimizing patient wellness. All of the above apply from the pediatric to the geriatric.

Work Environment

Nurses work in a variety of places and environments, beyond the hospital. They serve in clinics, schools, rehabilitation centers, outpatient and mental health facilities, ambulatory care centers, private physician clinics, medical offices, community centers, nursing homes, and patient's homes. And some places you may not immediately think of such as camps, homeless shelters, prisons, sporting events, and tourist destinations. Nurses even work at film studios. The work environment does play an integral role in the type of services offered. Also, nurses are always prepared to work long hours depending upon the emergency and environment and since patients get sick 24/7 then nurses also work weekends, holidays, and night shifts. And during these shifts the nurses spend most of their time walking, standing, lifting, and bending. On top of this many nurses are supposed to work closely with patients suffering from infectious disease. All in all, risky business.

Nurse Fashion

There are a handful of jobs where the clothing worn instantly reveals the person's profession. Police officers, firemen, members of the clergy, and soldiers fall into this

16. History of Nursing

category. As does the nurse. When we see someone in traditional nurse attire, a white dress and that characteristic cap, no introduction in necessary. In the film industry, having someone dressed in a uniform onscreen provides a no-cost backstory, something fiscally-minded producers like. As functional clothing the design of a nurse's uniform is centered around hygiene and identification. For her role in SF films the nurse in her uniform adds verisimilitude and a sense of authority. The mere presence of a nurse in the background of a scene adds such credibility.

The first nurse uniforms were derived from nun's habits which makes sense since nuns took care of sick and injured people. One of Florence Nightingale's first students at her school, Miss van Rensselaer, designed the original uniform—a floor length blue outfit with a white pinafore apron or smock over it and a frilly nurse's cap. The nurse uniform began to look more professional in order to distinguish trained nurses from those who were not. An all-white uniform became dominant during the early 20th century and continued through World War II. During the 1960s open necks began to appear, sleeves became shorter, and different style starched caps replaced the traditional cotton ones. By the late 1970s caps began to disappear from traditional hospital settings. From the 1990s emphasis has been on functionality, so scrubs of many colors, shapes, fabrics, and prints have became popular though some, primarily in the private sector, continue to wear the traditional white dress and cap. Initially, male nurses generally wore a white tunic and white pants and currently wear multi-colored uniforms or scrubs.

The main difference in the decades of nurse involvement in film, either minor or major, can readily be judged by the clothes they wear. Especially women's clothes whereas male nurses have essentially been the same during the same years. Each decade of film, from the silent era to current, has a slightly different nurse attire. From all white dresses, caps, and pants to multi-colored tops, now the norm in 21st century hospital environments. Shoes vary from mostly white during early cinema to all the colors of the rainbow in the 21 century.

One item nurses do not wear on the job are high heels, since walking, and occasional running (for emergencies), is a major part of the job. During any given shift nurses can walk up to 10 miles or even more, so foot comfort is essential.

Nurse Caps

There is a debate amongst nurses whether the nurse cap is a symbol of authority or servitude. Either way the nurse cap provides a finishing touch to the uniform and help identify a nurse's rank and grade. Initially, the white caps covered the entire head but became smaller and smaller over the decades. The first caps were a softer linen and during the early 20th century the caps began to be more starchy with pointed corners. After World War II some caps have color stripes around the rims.

The Nurse in Science Fiction Films

Nurse Jewelry

For the most part nurses are discouraged from wearing jewelry since it may get caught on patient skin, or interfere with their manipulations, not to mention serving as a possible distraction from their care activity. The only jewelry item that nurses do wear are watches.

The Films

Listing all SF films with nurses would be essentially a book unto itself so we will limit ourselves to a few examples from each decade. In many films the nurse is just seen in background, another body to populate the scene and to give some verisimilitude. In other films the nurse has a minor, though important role, while in others the nurse has a major substantive role. Though scenes of nurses may be brief, the number of films they are in is anything but.

What is interesting is that the number of films with nurses (admittedly, this is not an exhaustive list) significantly increases from the silent era and throughout the 1950s. In early cinema the nursing profession was still in its infancy and this was reflected in their scant appearances in films and when they did appear their roles were one-dimensional. If they weren't just standing there they usually held the proverbial tray in their hands. Once World War II began during the early 1940s the role of nurses was significantly propelled into prominence and this too was reflected in cinema. Films during the 1950s seem to have peaked with the appearance of nurses in just about every SF film produced. Each decade since then has had its share of nurses though not as prominent as the 1950s. However, the sophistication and work performed by nurses in SF films since the 1960s mirrors the advent of modern medicine technology (post–1960) in that the modern nurse now performs many of the functions of a pre–World War II physician. An example is the syringe injection which was always performed by a doctor pre–World War II and after the war nurses frequently were seen administering an injection.

Silent Films

The Hunchback of Notre Dame (1923)

This film takes place during Medieval times ("10 years before Columbus," c., 1482) and therefore way before the existence of nurses. Nevertheless, Esmeralda, though not a formal nurse, essentially performed the functions and duties of a nurse in providing help and support for the Hunchback while he was chained to the wheel. Quasimodo cried, "I thirst!" as Esmeralda walked by and she gave him something to

16. History of Nursing

drink, essentially taking care of his wellbeing. She also puts his shirt back on to cover his whipping scars. All the things a nurse would do in the normal care of their patients.

1930s

Dracula (1931)

Several nurses appear in this film and their roles range from backgrounder filler to major plot influencer. The first nurse is shown during a surgical demonstration in an auditorium, where she is assisting the surgeon. In one wide-view shot of the grounds at Seward's Sanitarium several nurses are seen with one in particular pushing a patient in a wheelchair. Just the simple presence of a nurse without any specific duties is a very common role in early cinema and this scene is one typical example.

Nurse Briggs has a more consequential role in the film. She was instructed to place wolfbane around Mina's room but after Mina objected, she obligingly removed the plant thinking she was helping Mina, but unfortunately, this act allowed Dracula to enter Mina's room. Also, Nurse Briggs opened the window that allowed Dracula to enter.

The Mummy (1932)

A home nurse takes care of Helen Grosvenor, the reincarnation of Ank-se-namun, while she is recovering from the attention of Ardeth Bey. The nurse, who makes two appearances, exits Helen's room and reports, "I've given her some bromide. She's asleep now." A bromide in the 19th and early 20th century was given as a sedative. The administration of medications ("meds") is one of the prime functions of the nurse.

In the 1931 *Dracula*, Nurse Briggs (Moon Carroll) allows Dracula (Bela Lugosi) entrance to Mina's room.

1940s

Black Friday (1940)

There is a nurse at a patient's bedside and she immediately gives the

The Nurse in Science Fiction Films

"I've given her some bromide," says a nurse in *The Mummy*. (Left to right, Zita Johann, unidentified actress, Kathleen Byron.)

arriving doctor a patient update, "Pulse 65. He's in a coma." Later, a nurse enters the patient's recovery room to inform the doctor that someone wants to see him. Another nurse, seen in the background, is exiting a room with a container in her hand. In SF cinema nurses are very frequently seen carrying or holding something, usually a tray.

Man Made Monster (1941)

In an a rare production nod, a "Nurse" is listed in the opening credits.

When she makes her appearance, the nurse, wearing a traditional white uniform and cap, says, "I can't make him stay in bed," referring to the uncooperative Dan McCormick. Nurses are seemingly always "fighting" with their patients in trying to get them to do things for their own good.

The Strange Case of Dr. Rx (1942)

At the Mayview Hospital there is a brief appearance of a nurse who has a tray with drink as she enters a patient's room. A second nurse is a home nurse seen in

16. History of Nursing

background preparing some sort of sample. Both nurses are wearing traditional white caps and uniforms.

Frankenstein Meets the Wolf Man (1943)

At Queen's Hospital, Ward B, we see the nurse inform the doctor, "He's conscious, Dr. Mannering. And talked. He insists upon sitting up." The nurse has a clipboard with the patient's information. The nurse is wearing a striped dress covered with a white smock and a white cap. Her shoes and nylons are a dark color.

As hospital rooms go the one Larry Talbot is in is a stark hospital ward with way too much wasted space. It should be noted that the patient's chart is attached to the wall over Talbot's bed which is unusual because charts are usually attached to the bed in case the bed gets moved so the chart goes with the bed and patient. As attached to the wall the chart could easily be separated from the patient if the bed were to be moved. In this ward is seen a second nurse sitting at a desk keeping watch on the patient. She too is wearing the same uniform as the first nurse.

Dr. Mannering says to Talbot, "The nurse tells me you are well enough to talk," indicating one of the key roles of nurses, informing the doctor of their patient's status. To follow up on this, Mannering later says to the nurse, "If there is any change call me immediately." Then the nurse puts Talbot back into bed and adjusts his pillow.

Later in the film the nurse says to Mannering, "I found this bandage here when I came in this morning. I didn't open the window. The patient must have done that himself." The nurse keeps the doctor updated and then follows his subsequent directions. Here Mannering instructs, "Better cover him up."

Later, Mannering says, "Get some more bandages, nurse. We'll renew the compress." The nurse is then seen bringing the proverbial tray with the compress materials.

When Talbot has a physical episode Mannering summons three male nurses to overpower Talbot and put a straightjacket on him, a job that would be difficult for most female nurses. These male nurses are wearing white pants and white front-buttoning shirts and dark shoes.

The Return of the Vampire (1943)

Early in the film we see a nurse nanny in charge of two children. Her uniform is a dress covered with a white smock and a white cap. Another nurse is at a patient bedside, assisting in his examination. Her uniform is similar to the nanny nurse with a white smock over a dress and white cap. Off-screen we hear her shouting, "Dr. Ainsley! Dr. Ainsley!"

Later, Lady Jane orders, "Call the nurse at once," demonstrating that nurses do not have to be physically present to make an impression. The mere mention of a nurse implies that competent, compassion care is needed.

The Nurse in Science Fiction Films

Captive Wild Woman (1943)

At the Crestview Sanitarium we see one nurse, with typical white uniform and cap, who can do everything. She brings patients to see the doctor, prepares the patients for examination, assists in the laboratory, and helps with operating nurse duties. This talented nurse also goes way above and beyond standard expectations of service, though not voluntarily, by "donating" her cerebrum to further "Dr. John Carradine's" brain experiments.

House of Dracula (1945)

In an interesting twist in this film nurse Nina is a hunchback. She wears the typical white uniform with a white smock, white hat, white nylons, and white shoes. Her fellow nurse, Miss Miliza Morelle also has a white uniform that buttons down the front, white hat, white shoes, and standard nylons. Nina helps with a blood transfusion and carries around the proverbial trays. Both nurses show concern for the well-being of patient Lawrence Talbot and try to take care of him. Both nurses are multi-talented. In addition to nurse duties they are also lab assistants and help Dr. Edlemann with his work, assisting in the operating room and serving as nurse anesthetists. The film allegedly takes place circa 1880 but this is difficult to accept for several reasons. The uniform of the nurses appears to be more 1940s-style than 1880s. Furthermore, Dr. Edlemann uses a sophisticated phone (not invented until the 1900s!) not to mention his 1930s microscope!

1950s

Note: of the many SF films throughout the 1950s just about all of them have a nurse on screen so we will have to limit ourselves to a few examples that are representative of the decade.

Abbott and Costello Meet the Invisible Man (1951)

Near the end of the film a nurse is seen standing bedside with the invisible man (Arthur Franz). The doctor instructs, "Reagent, nurse," and the nurse hands the doctor a syringe and a cotton ball (most likely loaded with alcohol as a sterilizer for the needle). At Lou's bedside is another nurse. Soon after that, Lou, rendered invisible from backwash blood from a transfusion, enters an elevator full of nurses, all decked out in the traditional white uniform.

The Day the Earth Stood Still (1951)

There is a brief scene that takes place at the Walter Reed Army Hospital in Bethesda, Maryland. A nurse is seen carrying a metal tray with a bottle and a few other items on her way into Klatuu's hospital room. Klatuu has escaped and the nurse looks around in amazement. She is wearing typical attire of the day. This is an example of the nurse not really having anything significant to do other than helping advance the plot. There is also a shot of Klatuu looking out of his hospital window at the grounds below and we see a nurse pushing a patient in a wheelchair, a common duty of many nurses, and a common scene in SF cinema.

Beast from 20,000 Fathoms (1953)

At Hartley Hospital a nurse opens a patient's room door to let in an army colonel. Another nurse brings the same patient a tray of food, engages in conversation to monitor the overall health of the patient, and then straightens out his bed sheets and fluffs pillow. Routine activities of a nurse, further demonstrated in the scene in Canada, where a nun is shown taking care of a patient and serving the same role as a nurse and caregiver.

Godzilla: King of the Monsters (1954)

After Godzilla has caused much mayhem there is a general scene of the wounded being brought to a hospital for care of their injuries. Many Japanese nurses are seen going about the business of triage. All of the nurses are seen with similar white dress uniforms but have two types of hats that distinguish their duties. On the front of the hats is a cross signifying their health care responsibilities.

The Quatermass X-Periment (1955)

Several nurses are seen around the Central Clinic and lab going about their normal duties. Nurses are seen walking hallways and pushing a patient's bed into an elevator. One nurse sees the first victim in an elevator. Also shown is a surgical nurse assisting an operation. These nurses are wearing a mid-calf dress with a white smock covering it with white cap and shoes.

X-The Unknown (1956)

Nurses are seen at the radiation department of a hospital examining patients. The nurse uniforms are different with two colors and a stylish hat. Nurses are in the background moving about without any particular purpose. In one scene a doctor and nurse engage in a kissing scene (more common than you might imagine!).

Not of This Earth (1957)

In a hospital setting we first see nurse Nadie at a desk greeting a patient and providing information. She is wearing the standard white uniform and cap with a color stripe around the rim. The nurse hangs a blood transfusion bottle and the patient says, "Are you a good one?" and she responds, "That's no question to ask a nurse" (indeed!).

Later a doctor gives Nadine orders for continued blood transfusions in a home care setting and she agrees. As a home care nurse for Paul Johnson, who's not of this earth, Nadie's primary job is to administer blood transfusions. She wears the same white uniform but no cap.

Later in the film two other nurses are seen interacting with a doctor and carrying out his orders that involve scheduling tests and working with blood samples. A nurse also unwittingly provides a major plot point by placing contaminated blood in storage that was later used in a fatal transfusion.

Attack of the 50 Foot Woman (1958)

A home nurse takes care of Mrs. Archer, the 50 foot woman. The nurse follows the doctor's orders, prepares a hypodermic needle injection, and sleeps in the same room as the patient. "Something has happened to Mrs. Archer!" exclaims the nurse when she sees the 50 foot woman for the first time. After the 50 foot woman wakes up and causes some havoc the doctor shouts, "Nurse, more morphine!"

Revenge of Frankenstein (1958)

The action takes place in the 19th century, circa 1860, mostly at a clinic ("Chirurgie"; German for surgery) that treats the poor. Madeline, the "Lady with the Basket," is a volunteer at the clinic and gives out soap (no takers), writing paper (again, no takers), and tobacco (many takers). She comforts a patient and distracts him with conversation. She wore a period dress covered with a white smock and no hat. Since Florence Nightingale established her school of nursing in 1860, contemporary with the time of the film, the Lady with the Basket may have heard of the work by Nightingale and was inspired to volunteer at the clinic.

Another male "nurse helper" at the clinic is carrying the proverbial tray with supplies and helps the doctor. This nurse is wearing a light blue smock. Later he is asked by the doctor to dress a wound, which is a fairly remarkable task for a nurse in 1860.

(Note: Many Hammer films are mostly Victorian period pieces with clothing/sets to match. Nursing in Victorian England was contemporary with Florence Nightingale. Were Hammer nurses in vogue with the practices of Nightingale? Or were these nurses pre–Florence therefore making the nursing science a stretch for the time? Florence established her nursing school in 1860 so the nurses shown in Victorian Hammer films most likely were trained at her school. Maybe Van Helsing or Victor Frankenstein were instructed by The Lady With the Lamp, Florence Nightingale, at St. Thomas Hospital in proper patient care.)

16. History of Nursing

The Hideous Sun Demon (1959)

After exposure to intense radiation, Dr. Gilbert McKenna, is taken to a hospital for observation and recovery. Two nurses admit McKenna into the hospital and begin to take his vital signs. Another nurse is in the background going through some files. A nurse enters McKenna's room and checks his chart. She also checks his pulse and talks to the patient to assess overall health. As happens way too frequently (more common than reported) the patient flirts with a blonde nurse.

As part of her duties a nurse pushes wheelchair-bound McKenna to an open area to get some fresh air and sun ... unfortunately, too much sun!

1960s

The Hypnotic Eye (1960)

A male nurse stands at a bedside holding an I.V. bottle (Note: the bottle is perhaps a little too low for an effective drip since gravity is necessary for proper function and a too low bottle will provide an inefficient fluid flow). Though a brief scene it is plausible simply because the male nurse is wearing a white uniform; his presence does nothing to further the plot since an I.V. stand would be just as, if not more, effective.

The Devil's Hand (1961)

Though shown only briefly two nurses appear in this film, both doing work in the background. They serve no real purpose other than portraying "medical professionals" doing their jobs. And this is fairly typical of SF films that show nurses in the background, they are essentially one-dimensional characters who help provide an image of a medically professional setting.

The Brain That Wouldn't Die (1962)

A surgical nurse is assisting the doctors in an operating room. She is monitoring the patient's pulse, anesthesia, and informs the doctors of the patient's status. The main character, Jan, is herself a nurse and becomes the main focus of the film as the brain that wouldn't die (a.k.a., "Jan in a pan").

X—The Man with the X-Ray Eyes (1963)

A nurse is seen walking the halls of a hospital. Another nurse sits at a desk in a small ward, apparently to keep direct contact with the patients, and escorts Dr. Xavier to a particular patient's bedside. These nurses are wearing the traditional white uniforms and caps. Later in the film a team of surgical nurses assist Dr. Xavier in an operation and these nurses are all gowned in standard attire for surgery.

Gorgon (1964)

At the Vandorf Medical Institution circa 1910, post–Victorian (although 20th century phones make an appearance) the nurses wear varying attire. One nurse sports a white apron and light blue top, no cap, and black shoes. A second nurse, wearing a dark blue dress with a white smock and also wearing a chef-like white hat, is seen taking care of a mental patient. The nurse, Carla (played by Barbara Shelley), is also Magera, who is the Gorgon.

Gidorah—The Three-Headed Monster (1965)

Several nurses serve no real purpose other than adding credibility to a medical environment. One of the nurses is seen holding the proverbial tray and does as she is directed by the doctor. All of these nurses are wearing the traditional white uniform and white cap.

Navy vs. the Night Monster (1966)

On a Navy base on the remote Gow Island are a group of Navy nurses. The nurses are stationed at the island infirmary. These Navy nurses do not wear the usual white outfit and cap but instead wear traditional Navy uniforms. Nurse Nora Hall (Mamie Van Doren) wears a two tone uniform distinguishing her from the other nurses. As part of her duties nurse Hall hangs IV fluids for a patient.

The Vulture (1967)

At a hospital a nurse prepares a hypodermic and gives it to a doctor for injection. The nurse is wearing a pinkish dress with a white smock over it, a wide black belt, and white cap. This nurse is notable since she is an early example of the shift away from the all-white uniform. Later the same nurse enters the patient's room holding the proverbial tray.

Scream and Scream Again (1969)

We see the same nurses several times in the film and it doesn't take long to learn that they are on the bad guys' team. One nurse is frequently shown holding a metal tray with instruments and is seen doing traditional nurse procedures and behaviors. The nurses wear a bit more stylish white dress and white cap.

16. History of Nursing

1970s

Blood Mania (1970)

Nurse Miss Turner takes care of a wealthy patient at his home. Attending to his medical needs, obtaining prescription meds, preparing and serving meals in her white dress and white cap even in the home. Later, she is seen taking a temperature with a thermometer, a common practice.

Asylum (1972)

In the "Lucy Comes to Stay" segment of this film, nurse Miss Higgins acts as a home nurse to closely monitor a patient just released from an asylum. Not a significant part per se, but again, the mere presence of a nurse does add much credibility.

Werewolf Woman (1976)

Nurses with a light blue dress covered with a white smock and a white cap walk through hospital halls, moving carts, and holding trays. A nurse in an all white uniform and white cap is seen starting an I.V. drip by inserting a syringe needle into the patient's forearm, a procedure performed by doctors in earlier films.

The Incredible Melting Man (1977)

A nurse is in the recovery room of the patient and she is wearing the traditional white uniform (front buttoning), white hat, white shoes, and white nylons. She takes orders from the attending doctor and appears to become the first victim as she is chased out of the hospital by the melting man.

Halloween (1978)

A brief scene at the beginning of the film has a nurse driving a doctor to meet Michael Myers. Unfortunately, while driving the nurse is chain smoking, in spite of being a health care provider who should know better. In one of the earlier representations of a non-traditional uniform (i.e., non-dress) the nurse is wearing a white pants uniform, white cap, white shoes, and a blue cape. She is attacked and pulled screaming from her car by the eponymous Mr. Myers.

1980s

Friday The 13th (1980)

At the end of the film, the lone survivor of Camp Crystal Lake is in a hospital room and a nurse gives her an injection in the gluteus muscle. The nurse is wearing

The Nurse in Science Fiction Films

a traditional white uniform and cap with a stripe around the rim. The nurse has no dialog, and is just there to suggest plot, closure, and patient care.

An Americal Werewolf in London (1981)

Nurse Alex Price (Jenny Agutier) works at an English hospital, where she's outfitted in a stylish hat, white uniform, and a wide black belt. In her uniform pockets are various tools of the trade such as scissors, clamps, and pens. Nurse Price performs standard tasks, such as examining patient David Kessler and looking at his chart. Later, she prepares an injection for the doctor and then restrains Kessler as the doctor administers the injection. Also, she force feeds Kessler saying, "you have to eat something."

Nurse Price also sees pediatric cases. Normally, there is a specialization so the same nurse would not have both children and adults to care for.

Re-Animator (1985)

At Miskatonic Medical School, Arkham, Massachusetts, emergency room nurses assist in the revival of a female patient early in the film as well as one later in the film. Also, several nurses are seen walking the halls of the hospital. One patient was taken to an emergency room for resuscitation and a few nurses helped. These nurses are attired in green scrubs and do not wear hats.

Lifeforce (1985)

We see a nurse in the background as she goes about her work filling a syringe for an injection. Though brief, it is representative of that common movie scene. After filling the syringe she places it on the proverbial metal tray and brings it to the patient's bedside. Not a word of dialog, but a scene in which the nurse, wearing a simple white uniform, and confidently executing a medical task, brings authenticity.

The Serpent & the Rainbow (1987)

Two nurses are seen in Haiti taking care of the main character in a hospital setting. One nurse appears to be a nun nurse since she is wearing a rosary around her neck and her white hat seems to more resemble a nun's hat than a nurse cap.

Bride of Re-Animator (1989)

In roles similar to that of the first *Re-Animator* movie, several nurses walk the halls of the Miskatonic Hospital, Arkham, Massachusetts. A nurse is seen hanging a plastic IV bag and taking doctor's orders. Gloria, a bride akin to that of Frankenstein, falls into disparate pieces after being rejected by her groom, in an event the doctor describes as "tissue rejection." Indeed. The nurses futilely attempt to assist Gloria, but they can't put her back together again.

1990s

Misery (1990)

Nurse Annie Wilkes rescues a writer from a car accident and claims to be his "number one fan." At first, Nurse Annie does take care of her patient and attends to his injuries and personal care, including shaving, something nurses do very frequently that isn't often depicted on film. After taking care of her patient she also inflicts severe pain and trauma followed up by the attendant care. An example of a nurse gone bad, Nurse Annie is distraught over the writer killing off a key character she likes in his latest book.

Darkman (1990)

At a burn unit in a hospital several nurses are seen doing unspecified work and serving mostly as bodies to fill the scene and add realism. Some of the nurses wear the traditional white uniform and cap whereas those directly working with a burn patient wear protective gowns, head covering, and face masks.

Outbreak (1995)

Most of the nurses in this film are seen in triage and emergency situations that, at times, can be quite chaotic, but true to nurse training, they remain calm and are there to take care of the patients. In the emergency room the nurses wear scrub green uniforms that are typical of the era. It should be noted that the patients are infected with a highly infectious disease ("Motaba virus" masquerading as ebola) and the nurses are there on the front lines putting their personal safety aside and dealing with the patients. Just like in real life.

Conclusions/Summary

The role of the nurse in early SF cinema was mostly background, a warm body to occupy screen space and move the plot along, and to serve as an authority figure and professional. Essentially, window dressing. After World War II, when the nursing profession catapulted into the institution we all know and understand, the appearance of nurses in films became more prevalent and for the most part moved beyond the

The Nurse in Science Fiction Films

window dressing days to become integral elements of film plots. The advances seen in the nursing profession since the beginning of cinema have been accurately represented in the contemporary films as each decade began. And like contemporary clothing, sets, and items such as cars, the overall attire and mannerisms of SF film nurses also define the time of film production.

When a nurse appears in pre–World War II films her primary role is one of a home nurse though there are numerous appearances, though brief, in hospital settings. Even so, their roles are of a simple nature such as walking around as background, carrying the proverbial tray, or just saying, "yes, doctor." After World War II, when the nursing profession became more visible to the general public this was also reflected in film plots. When SF film plots began to take place in hospital settings then nurses also became more prominent.

This is especially evident during the 1950s when many SF films involved doctors (mad, annoyed, or otherwise) along with their nursing staff. Since the 1960s the roles of nurses in SF films have become more sophisticated and reflects their more involved duties in real life. In addition to SF film plots the presence of nurses in television industry productions has helped catapult nurses into one of the most respected and sought after professions. Whether as backdrop or front-and-center in SF films the nurse does indeed take care of business and saves lives.

What will never change in film or fact, is that nurses tend to the human body and in the human spirit in ways that are hard to describe, but easy to convey with the universally recognized uniform and calm caring demeanor that commands respect, deserves appreciation and represents that competent authority that we automatically associate with the nurse.

Appendix
The Films

A Blind Bargain (1922)—Goldwyn Pictures
The Hands of Orlac (1924)—Berolina Film
The Hunchback of Notre Dame (1925)—Universal Studios
Dracula (1931)—Universal Studios
Frankenstein (1931)—Universal Studios
Dr. Jekyll & Mr. Hyde (1932)—Paramount Studios
Doctor X (1932)—Warner Bros.
Murders in the Rue Morgue (1932)—Universal Studios
Island of Lost Souls (1933)—Paramount Studios
The Invisible Man (1933)—Universal Studios
Bride of Frankenstein (1935)—Universal Studios
The Raven (1935)—Universal Studios
Mad Love (1935)—MGM Studios
The Werewolf of London (1935)—Universal Studios
The Man Who Lived Again (aka, *The Man Who Changed His Mind*; 1936)—Gaumont British
The Walking Dead (1936)—Warner Bros
The Devil Doll (1936)—MGM Studios
Son of Frankenstein (1939)—Universal Studios
The Ape (1940)—Monogram Studios
The Invisible Man Returns (1940)—Universal Studios
The Invisible Woman (1940)—Universal Studios
The Monster and the Girl (1941)—Paramount Studios
Man Made Monster (1941)—Universal Studios
Ghost of Frankenstein (1942)—Universal Studios
The Strange Case of DRx (1942)—Universal Studios
The Mad Monster (1942)—Producers Releasing Corp.
The Corpse Vanishes (1942)—Monogram
Frankenstein Meets the Wolf Man (1943)—Universal Studios
Undying Monster (1942)—20th Century Fox
Dr. Renault's Secret (1942)—20th Century Fox
The Ape Man (1943)—Monogram; Boris Karloff
Captive Wild Woman (1943)—Universal Studios
Jungle Woman (1944)—Universal Studios
House of Frankenstein (1944)—Universal Studios
The Invisible Man's Revenge (1944)—Universal Studios
The Monster Maker (1944)—Producers Releasing Corp.
Jungle Captive (1945)—Universal Studios
House of Dracula (1945)—Universal Studios
The Face of Marble (1946)—Monogram
The Beast with Five Fingers (1946)—Warner Bros.
Master Minds (1949)–Monogram
Bride of the Gorilla (1951)—Jack Broder Productions
Bela Lugosi Meets a Brooklyn Gorilla (1952)—Realart Pictures, Inc.
Beast from 20,000 Fathoms (1953)—Warner Bros.
Mesa of Lost Women (1953)—Howco Productions
The War of the Worlds (1953)—Paramount
Donovan's Brain (1953)—United Artists
Tarantula (1955)—Universal Studios
Bride of the Monster (1955)—Banner Pictures
The Werewolf (1956)–Columbia
The Monolith Monsters (1957)—Universal Studios
The Man Who Turned to Stone (1957)—Universal Studios
The Amazing Colossal Man (1957)—American International Pictures
The Beginning of the End (1957)—Republic Pictures
Giant from the Unknown (1957)—Astor Pictures
The Monster That Challenged the World (1957)—United Artists
The Man Without a Body (1957)—Filmplays, Ltd.
Invasion of the Saucer-Men (1957)—American International Pictures
She-Devil (1957)—20th Century Fox
Frankenstein's Daughter (1958)—Astor Pictures
The Colossus of New York (1958)—Paramount
Monster on the Campus (1958)—Universal-International
The Brain Eaters (1958)—American International Pictures

Appendix

Night of the Blood Beast (1958)—American International Pictures
Blood of the Vampire (1958)—Universal-International
The Alligator People (1959)—20th Century Fox
The Hideous Sun Demon (1959)—Pacific International Enterprises
Giant Gila Monster (1959)—McLendon-Rado Pictures
The Killer Shrews (1959)—McLendon-Rado Pictures
The Leech Woman (1959)—Universal-International
The Head (1959)—Universal Studios
The Wasp Woman (1959)—Filmgroup
The Angry Red Planet (1959)—American International Pictures
Caltiki. The Immortal Monster (1959)—Galatea Films
The Hands of Orlac (1960)—Continental Films
Konga (1961)—American International Pictures
The Curse of the Werewolf (1961)—Hammer Studios
Reptilicus (1961)—American International Pictures
The Witch's Mirror (1962); *El Espejo de la Bruja*—Cinematográfica
Hand of Death (1962)—20th century Fox
The Brain That Couldn't Die (1962)—American International Pictures
The Brain (1962)—British Lion / Columbia
Hands of a Stranger (1962)—Allied Artists
Atom Age Vampire (1963)—Film Selezione
Unearthly Stranger (1963)—Independent Artists
The Crawling Hand (1963)—Donald J. Hansen Enterprises
Madmen of Mandoras (1963) (aka, *They Saved Hitler's Brain*)—Paragon Films, Inc.
Flesh Eaters (1964)—Cinema Distributors of America
Last Man on Earth (1964)—American International Pictures
Frozen Alive (1964)—Feature Film Corporation of America
Mutiny in Outer Space (1965)—Wooner Brothers Pictures
War of the Gargantuas (1966)—Toho Studios
Astro-Zombies (1967)—Geneni Film Distributors
Corruption (1968)–Columbia
The Green Slime (1968)—MGM
A Clockwork Orange (1971)—Warner Brothers
The Andromeda Strain (1971)—Universal Studios
Brain of Blood (1971)—Independent International Pictures
The Incredible Two-Headed Transplant (1971)—American International Pictures
The Thing with Two Heads (1972)—American International Pictures
Horrors of the Blood Monsters (1972)—Independent International Pictures
The Creeping Flesh (1972)—Columbia
Soylent Green (1973)—MGM
The Beast Must Die (1974)—Amicus Productions
Logan's Run (1976)—MGM
Food of the Gods (1976)—American International Pictures
Werewolf Woman (1976)—Dailchi Film
Incredible Melting Man (1977)—American International Pictures
The Island of Dr. Moreau (1977)—American International Pictures
Humanoids from the Deep (1980)—New World Pictures
Saturn 3 (1980)—ITC Entertainment
Scanners (1980)—Avco-Embassy Pictures
The Empire Strikes Back (1980)—20th Century Fox
Outland (1981)—Warner Bros.
The Hand (1981)—Warner Bros.
The Howling (1981)—MGM
An American Werewolf in London (1981)—Universal Studios
The Thing (1982)—Universal Studios
Return of the Jedi (1983)—20th Century Fox
The Man with Two Brains (1983)—Warner Bros.
Dune (1984)—Universal Studios
Re-Animator (1984)—Empire International Pictures
The Serpent & the Rainbow (1987)—Universal Studios
Darkman (1990)—Universal Studios
Robocop 2 (1990)—Orion Pictures
The Addams Family (1991)—Paramount
Jurassic Park (1993)—Universal Studios
Deep Red (1994)—RHI Entertainment
Outbreak (1995)—Warner Bros.
Species (1995)—MGM
Screamers (1995)–Columbia
The Island of Dr. Moreau (1996)—New Line Cinema
Dark Angel (aka, *I Come in Peace*; 1997)—Triumph Releasing Corp
Matrix (1999)—Warner Bros.
Minority Report (2002)—20th Century Fox

Scary Monsters
Article Bibliography

1) Issue #76, "Living in the Lab of Luxury," pgs.. 29–35 (2010).
2) Issue #79, "The Maddest of the Mad Scientists," pgs. 66–70 (2011).
3) Issue #81, "Brains, Craniums, and Heads, Oh My!," pgs. 16–25 (2012).
4) Monster Memories #20, 2012 Yearbook, "The Labs of the Invisible Man," pgs. 20–26.
5) Issue #82, "The Hand in SF Cinema," pgs. 17–22 (2012).
6) Issue #83, "The Legacy of Doctor Moreau," pgs. 10–15 (2012).
7) Issue #84, "The Notebooks of Frankenstein," pgs. 37–46 (2012) [Rondo Award nominated as, "Best Article, 2012"].
8) Issue #85, "Dr. Gustav Niemann's Chalk Notes," pgs. 7–12 (2013) [Rondo Award nominated as, "Best Article, 2013"].
9) Issue #86, "Foods of the Gods," pgs. 27–36 (2013).
10) Issue #87, "Amazing Colossal Science," pgs. 19–20 (2013).
11) Issue #88, "Monkeying Around With Apes," pgs. 31–43 (2013).
12) Issue #89, "Dr. Van Helsing's Experiment," pgs. 6–8 (2013).
13) Issue #90, "The Microscope in Science Fiction Films," pgs. 50–62 (2014).
14) Issue #91, "The Laboratory of Dr. Septimus Pretorius," pgs. 67–77 (2014).
15) Issue #92, "Drugs in Science Fiction Cinema," pgs. 45–53 (2014).
16) Issue #93, "The Hairy Who are Scary," pgs. 48–60 (2014)
17) Issue #95, "The Spark of Life: Popular Science of Mrs. Mary Shelley," pgs. 7–15 (2015).
18) Issue #96, "Microbes in Science Fiction Films," pgs. 39–50 (2015).
19) Issue #97, "Hormones. The Scariest of Them All!," pgs. 29–43 (2015).
20) Issue #98, "Boris Karloff—The Walking FrankenDead," pgs. 6–16 (2015).
21) Issue #99, "The Skinny on Scary Skin," pgs. 41–57 (2015) [Rondo Award nominated as "Best Article, 2015"].
22) Issue #101, "The Nurse in Science Fiction Films," pgs. 14–23 (2016) [Rondo Award nominated as, "Best Article, 2016"].

General Bibliography

Animals Containing Human Material. http://www.acmedsci.ac.uk/policy/policy-projects/animals-containing-human-material/.
Bennett, John E., and Raphael Dolin. *Principles and Practice of Infectious Diseases.* 8th ed. New York: Elsevier, 2014.
Berg, A. Scott. *Lindbergh.* New York: Berkley Books, 1998.
Campbell, N.A., and J.B. Reece. *Biology.* 7th ed. New York: Pearson Benjamin Cummings, 2005.
Carroll, Karen C., Janet Butel, and Timothy Mietzner. *Medical Microbiology,* 27th ed. New York: McGraw-Hill, 2015.
Clark, D.P., and L.D. Russell. *Molecular Biology.* 3rd ed. St. Louis: Cache River Press, 2005.
Clay, Reginald S., and Thomas H. Court. *The History of the Microscope.* New York: Holland Press, 1982.
Dawkins, Richard. *The Selfish Gene.* Oxford: Oxford University Press, 1976
Forshier, Steve. *Essentials of Radiation Biology and Protection.* New York: Delmar Cengage Learning, 2002.
Glassy, Mark C. *The Biology of Science Fiction Cinema.* Jefferson, NC: McFarland, 2001.
Grove, Susan K., and Nancy Burns. *The Practice of Nursing Research: Appraisal, Synthesis, and Generation of Evidence.* 6th ed. New York: Elsevier, 2009.
Habif, Thomas B. *Clinical Dermatology.* 6th ed. New York: Elsevier, 2013.
Hall, John E. *Textbook of Medical Physiology.* 13th ed. New York: Elsevier, 2014.
Kandel, E.R., J.H. Schwartz, T.M. Jessell, S.A. Siegelbaum, and A.J. Hudspeth. *Principle of Neural Science.* New York: McGraw-Hill, 2013.
Karimi, Reza. *Biomedical and Pharmaceutical Sciences with Patient Care Correlations.* St. Louis: Jones & Bartlett Learning, 2015.
Kirby, David A. *Lab Coats in Hollywood: Science, Scientists, and Cinema.* Boston: MIT Press, 2011.
Kubler-Ross, Elizabeth. *On Death and Dying.* New York: Simon & Schuster, 1969.
Kumar, Vinay, Abdul K. Abbas, and Jon C. Aster. *Robbins Basic Pathology.* 9th ed. New York: Elsevier, 2014.
Montillo, Roseanne. *The Lady and Her Monsters: A Tale of Dissections, Real-Life Dr. Frankensteins, and the Creation of Mary Shelley's Masterpiece.* New York: Harper Collins, 2013.
Perkowitz, Sidney. *Hollywood Science: Movies, Science, and the End of the World.* New York: Columbia University Press, 2010.
Sadler, T.W. *Langman's Medical Embryology.* 13th ed. New York: Wolters Kluwer, 2014.
Seymour, Miranda. *Mary Shelley.* New York: Grove Press, 2000.
Shelley, Mary. *Frankenstein, or the Modern Prometheus.* London: Lackington, Hughes, Harding, Mayor & Jones, 1818.
Shlomo Melmed, Kenneth S. Polonsky, P. Reed Larsen, and Henry M. Kronenberg. *Williams Textbook of Endocrinology.* 12th ed. New York: Elsevier, 2011.
Warren, Bill. *Keep Watching the Skies!: American Science Fiction Movies of the Fifties, The 21st Century Edition.* Jefferson, NC: McFarland, 2010.
Weaver, Tom, Michael Brunas, and John Brunas. *Universal Horrors: The Studio's Classic Films, 1931–1946,* 2nd ed. Jefferson, NC: McFarland, 2007.
Wolf, Leonard. *The Annotated Frankenstein.* New York: Clarkson N. Potter, 1977.

Index

Abbott & Costello Meet Frankenstein 41
Abbott & Costello Meet the Invisible Man 230
Acid pentyl 118
Acromegaly 138, 139, 192, 193
Adler Theater 202
Agar, John 192
Age of Reason 67
Agrippa, Heinrich Cornelius 73
Agutier, Jenny 236
Alamogordo, NM 203
Alchemy 50, 73
Aldini, Giovanni 71
The Alligator People 130, 142
Alopecia 97
Alzheimer's disease 114, 131
The Amazing Colossal Man 7, 118, 132, 140, 145, 202, 206, 210, 217
Ambras Syndrome 99
American Red Cross 222
An American Werewolf in London 109, 236
Amphetamine 114
Amytal 116
The Andromeda Strain 120
Angel dust 119
The Angry Red Planet 162
Ank-se-namun 227
Ankers, Evelyn 37, 137
Antibodies 164
The Ape 134
The Ape Man 104, 133, 136
Aquanetta 137
Ardeth Bey 227
Arrakis 122
Arozamena, Eduardo 30
Asylum 235
The Atom 215
Atom Age Vampire 144, 145
Attack of the 50' Woman 232
Attention deficit disorder 114

Atwill, Lionel 171
Australopithecus afarensis 92
Axon 61
Ayahuasca 112

Bacta 121
Bactrian 215
Barry, Gene 160, 161
Barrymore, Lionel 190
Barton, Clarissa 222
Barty, Billy 56
Beast from 20,000 Fathoms 160, 231
The Beast Must Die 108
The Beginning of the End 140, 145, 193
Benson, Thomas 17
Best, James 195
Betel nut 112
Bettoia, Franca 164
Bible 156
Billings, Ted 36
Black Death 166, 188
Black Friday 227
Blaisdell, Paul 216, 217
Blake, Florence Guiness 222
Blanchard, Mari 141
Blarcy 124
A Blind Bargain 17
Blood Mania 235
Blood of the Vampire 22, 27
Bloom, John 172
Blue dreamers 120
Borel, Annik 108
Borgia, Lucretia 123
Boulder Dam 217
The Brain 175
The Brain That Wouldn't Die 174, 233
Brando, Marlon 178, 182
Brave New World 125
Breckinridge, Mary 222
Bride of Frankenstein 36, 41, 45, 54, 74, 84, 89, 158

Bride of Re-Animator 236
Brooks, Elizabeth 109
Browning, Tod 28
Buffalo Bill 222
Bufo marinus 123
Burke, Kathleen 134
Burke and Hare 74
Byron, Kathleen 228

Caffeine 112, 114
Calcium carbonate 42
Callistratus 22
Caltiki, the Immortal Monster 23
Camp Crystal Lake 235
Candida albicans 149
Cannary, Martha Jane 222
Captive Wild Woman 105, 137, 229
Carradine, John 105, 137, 159, 230
Carroll, Leo G. 191
Carroll, Moon 227
Carrel, Alexis 87
Casper's Law 81
Castle, Peggy 140, 193
Chaney, Lon, Jr. 35, 38, 103, 171
Chaney, Lon, Sr. 1
Churchill, Marguerite 86
Clarke, Arthur C. 5, 56
Clive, Colin 36, 45, 54, 66
A Clockwork Orange 119
Clostridium tetani 154
Club Med 44
Cobalt-60 142
Cocaine 114
Codeine 116
Coelacanth 106
Coffin, Tristram 133
Coleridge, Samuel Taylor 68
The Colossus of New York 172
Connery, Sean 121
Conservation biology 187

Index

Coogan, Jackie 139
Corday, Mara 192
The Corpse Vanishes 133, 136, 145
Corruption 144, 145
Cotton, Joseph 197
Coulouris, George 173
CPR 82
The Creeper 139
The Creeping Flesh 26
Crimean War 221
The Curse of Frankenstein 66
The Curse of the Werewolf 107
Cushing, Peter 26, 66, 144

Danieli, Emma 163
Dark Angel 124, 131
Darkman 237
Darwin, Charles 16
Da Vinci, Leonardo 72
Davis, Wade 123
Davy, Humphry 68, 72, 73
The Day the Earth Stood Still 230
The Day the World Ended 85
De Humani Corpus Fabricia 72
Dead Rising 125
Dee, John 74
Dendrite 61
Derma 28 144
Dermatology 208
Dern, Bruce 172
Descartes, Rene 130
Desert Rock 202
Destination Moon 6
The Devil Doll 190, 199, 215
The Devil's Hand 233
Dexedrine 114
Diabetes 143
Dick, Philip K. 117, 125
Digenerol 118, 119
Dix, Dorothea 222
DNA 3, 5, 15, 16, 34, 53, 63, 66, 67, 75, 106, 107, 129, 135, 137, 146, 149–151, 153, 163, 171, 177–180, 182, 183, 205, 206
Dr. Jekyll & Mr. Hyde 101, 134, 145
Doctor Moreau 177–183, 213, 217
Dr. Renault's Secret 135, 145
Donovan's Brain 64, 172, 174, 175
Dopamine 114, 116

Dracula 19, 28, 29, 78, 227
Drencrom 119
Dromedaries 215
Drosophila melanogaster 20, 141
Druktenis, Dennis 1, 127
Dune 122
"Duocane" 117

Ecstacy 113
Edelman, Franz 19
EEG 175
Egypt 73
Electricity 68, 69, 84
Electron microscope 24, 25, 26
The Empire Strikes Back 121
Endocrinology 67
Endorphins 131, 215
The Enlightenment 67
Ephemerol 120, 121
Ergot fungus 101
Erimin 116
Estrogen 136–139, 141, 144
Europe 67
Eustachius, Bartolommeo 72
Evers, Jason 174
Exodus 156

Famous Monsters of Filmland 5
Farber, Giovanni 13
Fermi, Enrico 202
Fielding, John 23
Fission 203
Flesh Eaters 164
Flunitrazepam 116
Food and Drug Administration (FDA) 113, 115, 117, 192
Food of the Gods 197
Ford, Grace 191
Frank, Horst 174
Frankenstein 6, 31, 34, 36, 51, 63, 71, 73, 78, 83, 85, 88, 89, 158, 170
Frankenstein, Henry 34
Frankenstein Meets the Wolf Man 33, 34, 38, 41, 71, 103, 229
Frankenstein's Daughter 118, 172
Franklin, Benjamin 68, 70
Franz, Arthur 106, 230
Freed, Bill 174
Friday the 13th 235
Frozen Alive 10

Frye, Dwight 170
Fungus 150
Fusion 203

Gadget 203
Galileo Galilei 13
Galvani, Luigi 69, 70, 71
Galvanism 68, 77, 213, 214
Garland, Beverly 142
Gates, Harvey 133
Ghost of Frankenstein 37, 41, 71, 83, 171
The Giant Gila Monster 142, 145
Gidorah—The Three-Headed Monster 234
Gila monster 142, 143
Glassy, Donna 219, 223
Globigerina 43
Godwin, Mary Wollstonecraft 66
Godwin, William 68
Godzilla 231
Goldilocks 206
Gordon, Bert I. 202
Gorgon 234
Gortner, Marjoe 198
Graham, Fred 142
Graves, Peter 140, 193
Gray, Coleen 143
The Green Slime 25, 26
Grosvenor, Helen 227
Growth hormone 136, 139, 140, 142, 144, 145
Guttenberg, Johannes 72
Gwenn, Edmund 84, 89

Hall, Henry 104
Halloween 235
Hamill, Mark 121
Hanna, Mark 202
Hardwicke, Cedric 37, 171
Harlan, Kenneth 133
Harrigan, William 117
Harrison, Harry 196
Harvard Medical School 78
Hatton, Rondo 139, 193
The Head 174
Heinlein, Robert A. 6
Heliobacter pylori 152
Heloderma suspectum 143
Hemoglobin 165
Hemsley, Estelle 143
Heroin 119, 125
Heston, Charlton 197
Heywood, Anne 175
The Hideous Sun Demon 233

Index

Hindbrain 59, 60
Hippocratic Oath 221
Hiroshima 203
Hirsuitism 99
HIV/AIDS 150, 154
Hologenome 151
Homo sapiens 92
Homunculus 50, 53, 55, 56, 58, 74
Hooke, Robert 13, 16
Hoover Dam 217
Hormone 67, 100, 114, 116, 126–139, 142–144, 181, 182, 192, 215, 216
Hoth 121
House of Dracula 7, 19, 151, 159, 230
House of Frankenstein 39, 41, 42, 43, 44, 45, 48, 64, 171
House of Horrors 139
Howard, John 104
The Howling 109
Hudson, William 211
Hull, Henry 102
Hull, Warren 86
Human growth hormone 131, 215
The Hunchback of Notre Dame 226
Hutton, Robert 173
Huxley, Aldous 125
Hydrocortisone 142
Hypertrichosis 99, 100
Hypnocil 125
The Hypnotic Eye 233

The Incredible Melting Man 235
The Incredible Shrinking Man 215
The Incredible Two-Headed Transplant 172
The Invisible Man 117
The Invisible Man Returns 117
The Island of Doctor Moreau 178
Island of Lost Souls 134, 145, 178

Jabba the Hut 122
Janssen, Zacharias 13
Johann, Zita 228
Jurassic Park 3, 4

Kalocin 120
Karloff, Boris 31, 39, 42, 45, 66, 77, 84, 88, 134, 171

Kamenniy-Ostrov 89
Kaun, Bernard 89
Keep Watching the Skies 202, 212
Khat 113
The Killer Shrews 195
King Lycaon 100
Kirkman, Robert 77
Klatuu 231
Knight, Victor 120
Knowles, Patrick 35
Koslec, Martin 164
Kruger, Otto 172

Lake Geneva 68, 74, 75
Lancaster, Burt 178, 182
Lanchester, Elsa 74
Langan, Glenn 140, 202, 217
Las Vegas 216, 217
The Last Man on Earth 163
Laughton, Charles 134, 178, 181
Launer, S. John 105
Lawrence, Sheldon 173
Lederberg, Joshua 147
The Leech Woman 130, 143, 145
Leeuwenhoek, Anton van 13, 14, 16, 148
Leith, Virginia 174
Leukocyte 18, 19
Leyden jar 69
LIFE magazine 121
Lifeforce 236
Lindbergh, Charles A. 87
Lindbergh heart 77, 87
Lippershey, Hans 13
Lloyd, Sue 144
Locher, Felix 118
Logan's Run 120
Lord Byron 68, 69, 75
Lovsky, Celia 197
LSD 101, 113, 124
Lugosi, Bela 17, 28, 29, 30, 78, 104, 134, 136, 171, 179, 227
Lumet, Baruch 195
Lupino, Ida 197, 198
Lupo, Alberto 144
Lycanthropy 100
Lynn, George 105

Macready, George 142
The Mad Monster 135, 145
Madmen of Mandoras 174
Mahoney, Mary Ezra 222
Make Room, Make Room 196

Malaria 150
Man Made Monster 228
The Man with Two Brains 175
The Man Without a Body 173
Manimal 178, 180, 182
Manning, Glen 118, 132, 199
Marca herbs 122
March, Frederic 101, 134
Mariana Trench 145
Marijuana 112
Mark, Michael 171
Martin, Ross 172
Martin, Steve 175
Massey, Ilona 35, 38
The Matrix 124
McCliam, John 197
McDowell, Malcom 119
McGoohan, Patrick 120
McKay, Wanda 139
Meeker, Ralph 198
Melange 122
Melanin 96
Melatonin 130
Melford, George 28
Merivale, John 2
Mesa of Lost Women 139, 145
Mescaline 113
Methamphetamine 114
Michelangelo 72
Middle Ages 153, 188
Mindjacks 125
Minority Report 124
Misery 237
Miskatonic Medical School 236
Mitochondria 25
Mohr, Gerald 162
Moloko plus 119
Monocane 117
The Monster and the Girl 171, 180
Monster Kid Scientist 5, 9
The Monster Maker 133, 138, 145
Monster on the Campus 22, 106
Morgan, Ralph 138
Morphine 116, 119, 232
Morphogens 59
Mother Gaia 186
Mount Tambora 74, 75
MRSA 157, 210
The Mummy 227
Murders in the Rue Morgue 133
Murphy, Donald 118, 172

Index

Muscle 120
Mutiny in Outer Space 20, 24, 26
Mycobacterium tuberculosis 152

Nagasaki 203
Naish, J. Carrol 39, 42, 138
Native American Indians 155
Naughton, David 109
Navy vs. the Night Monster 234
Neanderthal 107
Nembutal 116
Nerve growth factor 134, 136
Neuroin 124, 125
Neustadt Prison 43
New York 160, 221
Niemann, Gustav 42
Night of the Blood Beast 21
Night of the Lepus 145
Nightingale, Florence 221, 222, 225, 232
A Nightmare on Elm Street 3 125
Nipae 143
Nootropics 113
Normison 116
Norton, Edgar 17
Nostradamus 173, 174
Not of This Earth 231
Nuke 123
Nurse 219, 222, 224, 227
Nurse Education in Practice 223
Nursing Education Perspectives 223

Oak Ridge National Laboratory 162
Oland, Warner 102
Opium 113
Opuscula Anatomica 72
Ottiano, Rafaela 191
Oxycodone 118
Outbreak 165, 237
Outland 120, 121, 122

Paine, Thomas 68
Paracelsus 50, 74
Parasite 150
Parkinson's disease 114
PCP 119
Penicillin 111, 118
Peyote 113,
pH 28
physiology 91

phytohormones 131
Picardo, Robert 109
Pierce, Jck 103
Pineal gland 130, 131, 141, 145
Pituitary gland 131, 138, 142, 144, 145, 215
Placebo 115
Plague 155, 166
Planet of the Apes 105
Poe, Edgar Allan 78
Polidori, Dr. 75
Polydichloric euthimal 120, 121, 122
pons 59, 69
population biology 187
Pretorius, Septimus 49, 51, 52, 53, 55, 57, 58, 59, 62
Price, Vincent 163
Proceedings of the National Academy of Sciences 97
Psilocybin 113

Quasimodo 226
The Quatermass X-periment 231

Radiation 201, 204, 205, 207
Rains, Claude 117
Ramses II 156
Randolph, Jane 40
Rao, Umberto 163
Rathbone, Basil 17, 36, 158
RBC 17, 18, 19, 20, 23
Re-Animator 236
Red 123, 124
Reicher, Frank 171
Rembrandt 72
Reptilicus 23
Restoril 116
Return of the Jedi 122
The Return of the Vampire 229
Resuscitation 84
Revenge of Frankenstein 232
Ritalin 114
Ritch, Steve 105
Robinson, Ann 160
Robinson, Edgar G. 197
Robocop 2 123
Rockefeller Institute 87
Roosevelt, Theodore 113
Rubenstein, Anton Grigorevich 89

Saccharomyces cerevisiae 149
St. Thomas Hospital 221, 232

Sanders, Byron 164
Sanger, Margaret 222
Saturn 3 120
A Scanner Darkly 125
Scanners 120
Scary Monsters magazine 4, 7, 93, 127, 167
Scary Monsters 2009 Yearbook 1
Scream and Scream Again 234
Screamers 123
Seconal 116
Sequard, Charles Edouard Brown 127
Seracedia 139
The Serpent & the Rainbow 122, 236
Serum 114 119
She-Devil 20, 141, 145
Shelley, Barbara 234
Shelley, Mary 66, 67, 69, 70, 71, 72, 73, 74, 75, 76
Shelley, Percy Bysshe 68, 69, 75
Shrinkology 189, 190, 215, 216
Simon, Michel 174
Siodmak, Curt 175
Skywalker, Luke 121
Smallpox 155
Soma 125
Son of Frankenstein 17, 18, 19, 33, 36, 37, 41, 64, 158
Sorex soricidie 195
Soylent Green 196
Spacek, Sissy 175
Spacey, Kevin 165
Staphylococcus aureus 152, 154, 164
Star Trek 78, 125
Star Wars 121, 122
Sternhagen, Frances 121
Stevens, Onslow 19
Stevenson, Robert Louis 134
Stone, Christopher 109
Stone, Milburn 138
Strange, Glenn 135
The Strange Case of Dr. Rx 228
Strange Days 125
Substance-D 125
Sulfa hydral 118, 140, 214, 216
Surgery 167, 169
Sweat glands 98, 99
Synthemesc 119

Index

Tabula rasa 63
Tamblyn, Russ 25
Tarantella 139
Tarantula 191
Tekwar 125
Terminator 172
Terry, Philip 143, 171
Tetrodotoxin 123
Theophylline 114
Theraphosidae 139
Thesiger, Ernest 49, 50, 54
They Saved Hitler's Brain 174
Three Stooges 168
Thyroid-stimulating hormone 131
Tobey, Kenneth 161
Todd, Sally 172
Tower of London 173
"Trinity" 203
Transgenic 179–181
Trauma 169
Tremayne, Les 160
Trimorphonite 119

Undying Monster 104
Unearthly Stranger 119
United Nations 190
United States patent 4,666,425 173
University of California–San Diego 4, 167
University of Chicago 202
University of Edinburgh 223
"Urchin" 203

Van Eyck, Peter 175
Van Helsing 28, 29, 30
Van Rensselaer 225
Van Sloan, Edward 29, 34, 89, 170
Vellocet 119
Venus drug 125
Vesalius, Andreas 72
Vickers, Martha 138
Videohound 202
Villa Diodati 68, 75
Villar, Carlos 28
Virus 150, 163, 164, 165
Volta, Alessandro 69, 70
Voronoff, Serge 128
The Vulture 234

Wadsworth, William 68
Wald, Lillian 221
The Walking Dead 77, 84, 85, 87, 88, 89
Wallace, Dee 109
Walter Reed Army Hospital 230
Walters, Luana 133
Walthall, Henry B. 190
War of the Gargantuas 10, 24
The War of the Worlds 20, 153, 160, 161
Warren, Bill 202, 212
Wells, H.G. 178, 179, 197
The Werewolf 105
Werewolf Woman 108, 235
The Werewolf of London 102
Werewolf syndrome 99
Whitman, Walt 222
The Wizard of Oz 78, 167
Wolf Man 103, 106, 108
Wolfit, Donald 22
World War II 5, 113, 133, 203, 225, 226, 237, 238
Wrixon, Maris 135

X—The Man With the X-Ray Eyes 233
X—The Unknown 231

Yale School of Nursing 223
Yarborough, Barton 171
Yersinia pestis 156
Ygor 17

Zombinol 122, 123
Zombrex 125
Zucco, George 135, 171
Zygote 58